AF878968

Stochastic Analysis and Random Maps in Hilbert Space

Stochastic Analysis and Random Maps in Hilbert Space

A.A. Dorogovtsev

///VSP///

Utrecht, The Netherlands, 1994

VSP BV
P.O. Box 346
3700 AH Zeist
The Netherlands

© VSP BV 1994

First published in 1994

ISBN 90-6764-163-4

All rights reserved. No part of this publication may be reproduced, stored in a retrieval system, or transmitted in any form or by any means, electronic, mechanical, photocopying, recording or otherwise, without the prior permission of the copyright owner.

CIP-DATA KONINKLIJKE BIBLIOTHEEK, DEN HAAG

Dorogovtsev, A.A.

Stochastic analysis and random maps in Hilbert space /
A.A. Dorogovtsev. - Utrecht : VSP
With ref.
ISBN 90-6764-163-4 bound
NUGI 811
Subject headings: stochastic analysis / Hilbert space

Printed in The Netherlands by Koninklijke Wöhrmann bv, Zutphen.

Contents

Introduction

This book is devoted to stochastic operators in Hilbert space. A number of models in modern probability theory apply the notion of a stochastic operator in explicit or latent form. These models can be divided into two principal groups.

The first group includes such objects as random matrices, random linear operators as defined by Bharycha-Reid, differential and integral equations with random coefficients and kernels etc. In models of this type a random operator is taken as the set of deterministic operators which are indicated by points from some probability space. In this case there is a "good" operator (matrix, differential or integral operator etc.) for every ω .

The second group includes stochastic integrals with respect to random measures with independent values on disjoint sets and their generalizations as strong or weak random linear operators of Skorokhod. In this case there is no "good" deterministic operator for every ω, however the result of the action of the stochastic operator on elements of Hilbert space is observed.

In this book, objects from the second group in the Gaussian case are considered. Therefore it is useful to consider all random variables and elements as functionals from the Wiener process or its formal derivative, i.e. white noise.

The book consists of five chapters. The first chapter is devoted to stochastic calculus. Properties of the extended stochastic integral and stochastic derivative are considered. This introductory material is well-known and may be called L_2-theory because all random variables and elements to be considered have finite second moment. The main goal of the first chapter is to prepare the tools for solving stochastic equations.

The condition of the existence of second moment is very strong in many situations. So in this chapter stochastic calculus for random elements and variables which have no finite moments is built. The localization of the extended stochastic integral and stochastic derivative on a special set of random events is given. For certain generalized Wiener functionals, in the sense of Hida, ordinary random variables which coincide with them are found.

Stochastic calculus of white noise is closely related to analysis in infinitely-dimensional spaces. For this reason chapter 1 contains some statements about Sobolev spaces on open sets with Gaussian measure in Hilbert space.

In the second chapter the structure of stochastic operators, mainly the structure of Gaussian strong linear operators, is studied.

For a certain class of such operators, i.e. those cases when the operator is a set of deterministic bounded operators indicated by the points of probability space, Ω is described. For a general stochastic operator, the existence of representation as a sum of multiple stochastic integrals will be proved.

In chapter 3 the definition of the action of stochastic operator on random elements is considered. As has been mentioned above, in particular cases the stochastic operator is the stochastic integral. So in these cases the stochastic integral for the random integrand must be defined. In this book the extended stochastic integral is used as a tool. Another method of approach to the definition of the action of a stochastic linear operator on random elements is based on the symmetric stochastic integral. The relationship of these two methods is considered. In particular a generalized Fourier transform is constructed for the generalized white noise functionals. This transform is used for the comparison of the extended stochastic integral with the symmetric stochastic integral. The localization from chapter 1 provides the possibility of determining the action of a random operator on a random element which has no finite moments.

Chapter 4 is devoted to the mathematical models in which the notions of stochastic calculus arise. Local time intersection for the families of Markov processes, random statistics and systems with random values of particles is considered.

In chapter 5 the equation with random operators is considered. Theorems of the existence and uniqueness of solutions and the smoothness of solution in particular cases are proved.

Finally, I'd like to express my gratitude to academicians Anatoly Vladimirovich Skorokhod and Yury Lvovich Daletsky for useful discussions and questions.

Chapter 1

Stochastic calculus

1.1 Preliminaries (L_2-theory)

This section is devoted to the definitions of the extended stochastic integral and stochastic derivative and to their properties in the case when all random elements and variables, which are considered, have finite second moment.

Nowadays there are many excellent articles devoted to this subject. They are mentioned in the reference list. He who wants to study stochastic calculus in detail, can read the pioneering works of M. Hitsuda, A. V. Skorokhod and Yu. L. Daletsky and N. V. Paramonova or such surveys as the articles of T. Sekiguchy and Y. Shiota, E. Pardoux and D. Nualart.

This section contains well-known results and new facts which will be useful later. The statements are formulated in a convenient manner in order to provide the following exposition.

Let $(\Omega, \mathcal{F}, \mathrm{P})$ be a probability space and H be a real separable Hilbert space with inner product $(\cdot\,\cdot)$ and norm $\|\cdot\|$.

Let ξ denote the generalized Gaussian random element in H which has zero mean and identical correlation operator. In other words ξ is a linear map which maps elements of H into the set of Gaussian random variables and has the following property:

$$\forall\, \varphi \in H: \quad E(\xi, \varphi) = 0, \quad E(\xi, \varphi)^2 = \|\varphi\|^2.$$

The following examples will be very important for future reference.

Example 1.1.1. *Finite dimensional case.* If $\dim H = n$ ($n \in N$), i.e. $H = R^n$, then the generalized Gaussian random element ξ is a random Gaussian vector $(\xi_1, \ldots, \xi_n)$ such that

$$\forall\, i, j = 1, \ldots, n: \quad E\xi_i\xi_j = \delta_{ij}, \quad E\xi_i = 0.$$

Now

$$\forall\, \varphi \in H = R^n,\ \varphi = (\varphi_1, \ldots, \varphi_n): \quad (\xi, \varphi) = \sum_{j=1}^{n} \xi_j \varphi_j.$$

Example 1.1.2. *Gaussian random measure with independent values on disjoint sets.* Let (X, A, μ) be a measurable space with probability measure μ on σ-algebra

A. Consider a Gaussian random measure γ on A such that

$$\forall\, A, B \in A: \quad E\gamma(A) = 0, \quad E\gamma(A)\gamma(B) = \mu(A \cap B).$$

Then for every $\varphi \in L_2(X, A, \mu)$ the stochastic integral $\int_X \varphi(\theta)\gamma(d\theta)$ can be defined in the usual way. The value of this integral is a Gaussian random variable with zero mean and

$$E\Big(\int_X \varphi(\theta)\gamma(d\theta)\Big)^2 = \int_X \varphi^2(\theta)\mu(d\theta).$$

So ξ in $L_2(X, A, \mu)$ can be defined

$$\forall\, \varphi \in L_2(X, A, \mu): \quad (\xi, \varphi) := \int_X \varphi(\theta)\gamma(d\theta).$$

Example 1.1.3. *Measurable linear functionals.* Let the probability space $(\Omega, \mathcal{F}, \mathrm{P})$ be realized now as a separable Banach space B with Gaussian measure ν on Borel subsets of B. Let ν have a zero mean. Then, for every linear continuous functional $f \in B^*$, the distribution of f as a random variable is Gaussian and

$$Ef^2 = \int_B f^2(\theta)\nu(d\theta) = |f|^2,$$

where $|\cdot|$ is the norm of H-reproducing kernel for ν [1]. It is known that, for every element $\varphi \in H$, function $\varphi : B \to R$ is correctly determined as a limit of functions $f : B \to R$, $f \in B^*$ and has a Gaussian distribution with zero mean and variance $|\varphi|^2$. So the correspondence ξ which maps every element φ of H into a function (measurable linear functional) $\varphi : B \to R$ is a generalised random Gaussian element in H with zero mean and identical correlation operator.

For most of this book all random variables, elements, functions etc. will be measurable concerning a certain generalised Gaussian random element ξ (called further *white noise* in H). The L_2-calculus of the white noise functionals is discussed in this section. So $\mathcal{F} = \sigma(\xi)$ is taken. The main instrument of exposition now is the Wiener–Ito expansion. It will be spoken of in abstract form and in the particular cases of examples 1.1.1–1.1.3.

Let $H_k, k \geq 1$ be the space of k-linear symmetric Hilbert–Schmidt forms on H. Define the inner product and norm in H_k in the usual way:

$$\forall\, T_k, S_k \in H_k: \quad (T_k, S_k)_k := \sum_{t_1 \ldots t_k = 1}^{\infty} T_k(e_{t_1}, \ldots, e_{t_k}) S_k(e_{t_1}, \ldots, e_{t_k}),$$

$$\|T_k\|_k^2 = (T_k, T_k)_k \ ,$$

where $\{e_t;\ t \geq 1\}$ is an orthonormal basis in H. This definition is independent from the choice of the basis $\{e_t;\ t \geq 1\}$.

In other words, H_k is a symmetric part of the tensor product $H^{\otimes k}$. For $k = 0$ take $H_0 = R$. Let $Q_k \in H_k$, $k \geq 1$. In the basis $\{e_t;\ t \geq 1\}$, Q_k is defined by the matrix of the coefficients $\{q_{t_1 \ldots t_k};\ t_1, \ldots, t_k \geq 1\}$. Note that

$$\sum_{t_1 \ldots t_k = 1}^{\infty} q_{t_1 \ldots t_k}^2 = \|Q\|_k^2.$$

Define

$$Q_k(\xi,\dots,\xi) := \sum_{t_1\dots t_k=1}^{\infty} q_{t_1\dots t_k}(\xi,e_{t_1}) * \dots * (\xi,e_{t_k}),$$

where $(\xi,e_{t_1}) * \dots * (\xi,e_{t_k})$ is obtained from $(\xi,e_{t_1})\cdots(\xi,e_{t_k})$ in which $(\xi,e_t)^r$ is substituted by $\bar{H}_r((\xi,e_t))$ ($\bar{H}_r$ is the r-th Hermite polynomial). For example,

$$(\xi,e_1) * (\xi,e_3) * (\xi,e_1) = [(\xi,e_1)^2 - 1]\cdot(\xi,e_3).$$

It is easy to check that

1) $\forall\, k \geq 1\ \forall\, Q_k\,,\ T_k \in H_k$:

$$EQ_k(\xi,\dots,\xi) = 0,$$

$$EQ_k(\xi,\dots,\xi)\, T_k(\xi,\dots,\xi) = k!\,(Q_k,T_k)_k\,,$$

2) $\forall\, m > n \geq 0\ \ \forall Q_m \in H_m,\ T_n \in H_n$:

$$EQ_m(\xi,\dots,\xi)\cdot T_n(\xi,\dots,\xi) = 0.$$

The value of $Q_k(\xi,\dots,\xi)$ does not depend on the choice of basis $\{e_t;\ t \geq 1\}$.

The main property of $\{Q_k(\xi,\dots,\xi);\ Q_k \in H_k,\ k \geq 1\}$ is following (see [2]). Every random variable $\eta \in L_2(\Omega,\mathcal{F},\mathrm{P})$ can be developed in the series

$$\eta = \sum_{k=0}^{\infty} Q_k(\xi,\dots,\xi). \tag{1.1.1}$$

Sketch of the proof. Denote by $\mathcal{F}_n = \sigma((\xi,e_1),\dots,(\xi,e_n)),\ n \geq 1,\ \eta_n = E(\eta/\mathcal{F}_n)$. Then $\mathcal{F} = \bigvee_{n=1}^{\infty}\mathcal{F}_n$ and

$$E(\eta - \eta_n)^2 \to 0, \quad n \to \infty.$$

For every $n \geq 1$, η_n is a function from $(\xi,e_1),\dots,(\xi,e_n)$, and the existence of (1.1.1) for η_n follows from the fact that polynomials are complete in L_2 on R^n with Gaussian measure. Now the proof can be obtained using the properties 1) and 2) and the equality

$$E\left(\sum_{k=0}^{\infty} T_k(\xi,\dots,\xi)\right)^2 = \sum_{k=0}^{\infty} k!\ \|T_k\|_k^2\,.$$

Series (1.1.1) is called the Ito–Wiener expansion of η.

Consider examples.

Example 1.1.1′. In the case $H = R^n$, as has been mentioned above, the random variables $Q_k(\xi,\dots,\xi);\ Q_k \in H_k;\ k \geq 0$ are n-dimensional Hermite polynomials.

Example 1.1.2′. In this situation, for each $k \geq 1$, space H_k is a symmetric part of $L_2(X \times \dots \times X,\ \bigotimes_{j=1}^{k}\mu)$. So, for every $Q_k \in H_k$, there exists the unique symmetric kernel $q_k : X \times \dots \times X \to R$ such that:

$\forall\, \varphi_1,\dots,\varphi_k \in L_2(X,A,\mu)$:

$$Q_k(\varphi_1,\dots,\varphi_k) = \int \overset{k}{\dots} \int_X q_k(\theta_1,\dots,\theta_k)\,\varphi_1(\theta_1)\dots\varphi_k(\theta_k)\cdot\mu(d\theta_1)\cdot\ldots\cdot\mu(d\theta_k).$$

The random variable $Q_k(\xi, \ldots, \xi)$ is called a multiple integral and sometimes it is denoted by

$$\int \overset{k}{\cdots} \int_X q_k(\theta_1, \ldots, \theta_k)\gamma(d\theta_1) \ldots \gamma(d\theta_k).$$

When (X, A, μ) is $([0,1], \mathcal{B}, \lambda)$, where λ is the Lebesgue measure on the Borel σ-algebra $\mathcal{B}$ of $[0,1]$ and $\nu(d\theta) = dw(\theta)$, where w is the Wiener process, these multiple integrals are well-known Ito–Wiener integrals.

Sketch of the proof. Let $\varphi \in L_2([0,1])$, $\|\varphi\| = 1$, $\operatorname{esssup}|\varphi| < +\infty$. Consider the stochastic process

$$x(t) = \exp\left\{a \int_0^t \varphi(\tau)dw(\tau) - \frac{1}{2}a^2 \int_0^t \varphi^2 \tau d\tau\right\}, \quad t \in [0,1], \quad a \in R.$$

It is known that

$$dx(t) = x(t)\, a\varphi(t)\, dw(t).$$

Hence

$$x(t) = \sum_{k=0}^{\infty} a^k \int_{\Delta_k(t)} \varphi(s_1) \ldots \varphi(s_k)\, dw(s_1) \ldots dw(s_k),$$

where $\Delta_k(t) = 0 \le s_1 \le \ldots \le s_k \le t$, $k \ge 1$. From the other side

$$x(1) = \sum_{k=0}^{\infty} a^k \frac{1}{k!} \bar{H}_k \left(\int_0^1 \varphi(\tau)dw(\tau)\right).$$

So

$$\forall\, k \ge 1: \quad \int_{\Delta_k(1)} \varphi(s_1) \ldots \varphi(s_k)\, dw(s_1) \ldots dw(s_k) = \frac{1}{k!} \bar{H}_k \left(\int_0^1 \varphi(\tau)\, dw(\tau)\right).$$

Using this equality and taking the limit from step kernels, it can be obtained that

$$Q_k(\xi, \ldots, \xi) = k! \int_{\Delta_k(1)} q_k(s_1, \ldots, s_k)\, dw(s_1) \ldots dw(s_k),$$

for every $Q_k \in H_k, k \ge 1$.

Example 1.1.3′. Space B^* is dense in H. Hence for every $Q_k \in H_k$, $k \ge 1$, the random variable $Q_k(\xi, \ldots, \xi)$ is defined by the function $\tilde{Q}_k : B \to R$ which is a ν-limit of the sequence of usual polynomials of k-th degree on B. So $\tilde{Q}_k$ is a measurable polynomial with respect to measure ν.

Now consider the new important notion — stochastic derivative. Begin from the situation of example 1.1.3. The usual definition of the derivative for the function $f : B \to R$ is based on consideration of the ratio

$$f_{\Delta t}(\omega, h) = \frac{f(\omega + \Delta t h) - f(\omega)}{\Delta t}, \quad \omega, h \in B.$$

If the function f is differentiable at every point of the space B, then

$$\forall\, \omega \in B: \quad \lim_{\Delta t \to 0} f_{\Delta t}(\omega, h) = \langle f'(\omega), h\rangle,$$

where f' is the derivative of f. Hence, for the random variable f, there exists

$$\nu - \lim_{\Delta t \to 0} f(\cdot, h).$$

This fact provides the possibility of defining the derivative of a random variable, but there are two difficulties using this method.

Firstly, the random variable $f_{\Delta t}(\cdot, h)$ can be defined correctly for all $h \in B$ only for a small class of functions f. The cause of this lies in the properties of measure ν. Shifts on elements from H are only admissible for ν [1] and give measures which are equivalent to ν. So the random variable $f_{\Delta t}(\cdot, h)$ can be correctly defined only for $h \in H$ in the general case.

Secondly, the notion based on the limit of the ratio is closely connected with the situation of the example and may not be used in other situations. To overcome these mentioned difficulties, only derivatives along H can be considered and first the multiple Hilbert–Schmidt forms from ξ should be investigated. It can be easily checked that, for every h and every $Q_k \in H_k$, the ratio $Q_{k,\Delta t}(\cdot, h)$ is correctly defined and has its L_2-limit under $\Delta t \to 0$:

$$E(Q_{k,\Delta t}(\cdot, h) - kQ_k(h, \xi, \dots, \xi))^2 \to 0, \quad \Delta t \to 0.$$

Therefore, for multiple forms from ξ, the following definition is obtained.

Definition 1.1.1. Stochastic derivative of $Q_k(\xi, \dots, \xi)$ for $Q_k \in H_k,\ k \geq 1$ is a random element $DQ_k(\xi, \dots, \xi)$ in H such that

$$\forall\, h \in H : \quad (DQ_k(\xi, \dots, \xi), h) = k\ Q_k(h, \xi, \dots, \xi) \ \ (\mathrm{mod}\ P).$$

This definition is based on the usual definition of the derivative, but does not depend on the situation of example 1.1.3. The stochastic derivative for the random variable $\eta \in L_2(\Omega, \mathcal{F}, \mathrm{P})$ can now be defined with the help of the Ito–Wiener expansion.

Definition 1.1.2. The random variable $\eta \in L_2(\Omega, \mathcal{F}, \mathrm{P})$ with the Ito–Wiener expansion (1.1.1) is stochastically differentiable if the series

$$\sum_{k=0}^{\infty} DQ_k(\xi, \dots, \xi) \tag{1.1.2}$$

converges on the square mean in H. The sum of (1.1.2) is called a stochastic derivative of η and is denoted by $D\eta$.

Before studying the properties of a stochastic derivative, D, consider some examples.

Example 1.1.4. Consider $H = R$. Then ξ is a usual Gaussian random variable with a zero mean and variance 1. Multiple Hilbert–Schmidt forms from ξ are Hermite polynomials, and

$$\forall\ k \geq 0 : \quad DH_k(\xi) = H'_k(\xi) = kH_{k-1}(\xi),$$

where the symbol " $'$ " denotes a usual derivative. Therefore, the random variable $\alpha = f(\xi) \in L_2(\Omega, \mathcal{F}, P)$ is stochastically differentiable if, and only if, f has a derivative almost everywhere on R, is absolutely continuous and $f'(\xi) \in L_2(\Omega, \mathcal{F}, P)$. Under this condition $Df(\xi) = f'(\xi)$.

Example 1.1.5. Case $H = R^n$. Now $\xi = (\xi_1, \ldots, \xi_n)$, where $\xi_1, \ldots, \xi_n$ are independent identically distributed Gaussian random variables with zero mean and variance 1. Multiple Hilbert–Schmidt forms from ξ are multi-dimensional Hermite polynomials from $\xi_1, \ldots, \xi_n$ and, as in example 1.1.4, a stochastic derivative for them is a usual gradient. For example, if $n = 2$, $k = 2$, and form Q_2 is determined by matrix

$$Q_2 = \begin{pmatrix} 0 & 1 \\ 1 & 0 \end{pmatrix}$$

in the natural basis $e_1 = (1;0)$, $e_2 = (0;1)$, then

$$Q_2(\xi, \xi) = 2\xi_1 \cdot \xi_2,$$

$$DQ_2(\xi_1, \xi_2) = 2Q_2(\cdot, \xi) = 2\xi_1 e_2 + 2\xi_2 e_1 = \nabla \tilde{Q}(\xi_1, \xi_2),$$

where $\tilde{Q}(x_1, x_2) = 2x_1 \cdot x_2$, $(x_1, x_2) \in R^2$.

Hence, definition 1.1.2 implies that the space of such functions $f : R^n \to R$, for which $f(\xi_1, \ldots, \xi_n)$ is stochastically differentiable is Sobolev space $W_2^1(R^n, \mu)$, where μ is a distribution of the random vector $(\xi_1, \ldots, \xi_n)$.

Example 1.1.3″. It follows from example 1.1.3′ that, to describe these functions f on B for which the random variable $f(\cdot)$ is stochastically differentiable, a derivative along H must be used.

Definition 1.1.3. Function $f : B \to R$ is differentiable along H at the point $\theta \in B$ if

$$\forall\, h \in H: \quad f(\theta + h) = (\varphi, h) + r(h) + f(\theta),$$

where $\varphi \in H$, $r(h) = o(\|h\|)$, $h \to 0$ (in the H-norm). φ is a derivative of f at x along H.

Now the space of stochastically differentiable functions is the completed family $\mathcal{P}$ of all polynomials in the Sobolev norm

$$\forall\ g \in \mathcal{P}: \quad \|g\|^2 = \int_B g^2\, d\nu + \int_B \|g'_H\|^2\, d\nu.$$

The space of all stochastically differentiable random variables will be further denoted by W^1, where W^1 is a Hilbert space with the inner product of such kind:

$$\forall\, f, g \in W^1: \quad (f, g) = Ef \cdot g + E(Df, Dg)_1.$$

A further two generalisations of definition 1.1.2 will be necessary. These generalisations are higher stochastic derivatives and stochastic derivatives of random elements in a Hilbert space. Begin with the second type of derivative.

Let $\mathcal{H}$ be a separable Hilbert space with inner product $(\cdot,\cdot)_{\mathcal{H}}$ and norm $\|\cdot\|_{\mathcal{H}}$ and x a random element in $\mathcal{H}$ with finite second moment of the norm. Denote by $\{e_k;\ k \geq 1\}$ an orthonormal basis in $\mathcal{H}$. Then, for every k, the random variable $(x, e_k)_{\mathcal{H}}$ has the Ito–Wiener expansion

$$(x, e_k)_{\mathcal{H}} = \sum_{j=0}^{\infty} Q_j(\xi, \ldots, \xi; e_k).$$

It can be easily checked that

$$\sum_{k=1}^{\infty} \|Q_j(\dots;e_k)\|_j^2 \le \frac{1}{j!} E\|x\|_{\mathcal{H}}^2, \quad j \ge 0.$$

So the sum

$$\sum_{k=1}^{\infty} Q_j(\dots;e_k)\, e_k = Q_j(\dots)$$

exists and is a symmetric j-linear Hilbert–Schmidt form on H. Space of all such forms is denoted by $H_j(H)$. Hence

$$x = \sum_{j=0}^{\infty} Q_j(\xi,\dots,\xi), \tag{1.1.3}$$

where

$$Q_j(\xi,\dots,\xi) := \sum_{k=1}^{\infty} Q_j(\xi,\dots,\xi;e_k)e_k$$

(definition is correct and does not depend on the choice of a basis). Equality (1.1.3) is called the Ito–Wiener expansion of a random element x. Members of (1.1.3) are orthogonal elements in the space $L_2(\Omega, P, H)$, the space of all random elements in H with a finite second moment and the usual inner product.

Definition 1.1.4. For $j \ge 1, Q_j \in H_j(\mathcal{H})$, the stochastic derivative of $Q(\xi,\dots,\xi)$ is a random element in $H_1(H)$ for which such an equality occurs

$$\forall\, \varphi \in H: \quad DQ_j(\xi,\dots,\xi)(\varphi) = jQ_j(\varphi,\xi,\dots,\xi).$$

Definition 1.1.5. Random element $x \in L_2(\Omega, P, H)$ with the Ito–Wiener expansion (1.1.3) has the stochastic derivative Dx if the series

$$Dx = \sum_{j=0}^{\infty} DQ_j(\xi,\dots,\xi)$$

converges in $L_2(\Omega, P, H_1(H))$.

Now consider higher stochastic derivatives with the help of induction. Assume that, for the random element $x \in L_2(\Omega, P, H)$, the k-th stochastic derivative is determined as the random element $D^k x$ in $H_k(H)$ with a finite second moment.

Definition 1.1.6. The random element x has the $k+1$-th stochastic derivative $D^{k+1}x$ if the random element $D^k x$ in $H_k(H)$ has a stochastic derivative. In this case

$$D^{k+1}x := D(D^k x).$$

Example 1.1.6. For the random variable $\bar{H}_n((\varphi,\xi))$, $\varphi \in H$, $\|\varphi\| = 1$, $k \le n$,

$$D^k\left(\bar{H}_n((\varphi,\xi))\right) = \frac{n!}{(n-k)!}\, \bar{H}_{n-k}((\varphi,\xi)) \cdot \varphi \otimes \dots \otimes \varphi.$$

Further examples will be considered after studying the properties of a stochastic derivative. It can be stated for the case of random variables but can easily be adapted for random elements.

Property 1.1.1. D, with domain of definition W^1, is a closed linear map from $L_2(\Omega, \mathcal{F}, \mathrm{P})$ into $L_2(\Omega, \mathrm{P}, H)$.

Proof. Linearity is trivial. To prove that D is closed, it is sufficient to use the uniqueness of the Ito–Wiener expansion and the definition of D .

Property 1.1.2. Let L be the subspace of H and the random variable $\alpha \in W^1$ be measurable with respect to $\mathcal{F}_L = \sigma(\{(\xi, \varphi)\,; \varphi \in L\})$. Then $D\alpha$ is also measurable with respect to $\mathcal{F}_L$ and

$$\forall\, h \in L^\perp : \quad (D\alpha, h) = 0 \quad (\mathrm{mod}\ \mathrm{P}).$$

Proof. The measurability of α with respect to $\mathcal{F}_L$ makes it possible to organise the Ito–Wiener expansion thus

$$\alpha = \sum_{k=0}^{\infty} T_k(\xi_L, \ldots, \xi_L), \tag{1.1.4}$$

where $T_k \in L_k,\ k \geq 0$, and ξ_L is a generalised Gaussian random element in subspace L, which is determined by the equality

$$\forall\, \psi \in L : \quad (\xi_L,\ \psi) = (\xi, \psi).$$

Define new forms $\tilde{T}_k \in H_k,\ k \geq 0$, in such a way:

$$\forall\, \varphi_1, \ldots, \varphi_k \in H : \quad \tilde{T}_k(\varphi_1, \ldots, \varphi_k) = T_k(P_L\varphi_1, \ldots, P_L\varphi_k),$$

where P_L is a projector on L. Then

$$\forall\, k \geq 0 : \quad \tilde{T}_k(\xi, \ldots, \xi) = T_k(\xi_L, \ldots, \xi_L).$$

Hence, from (1.1.4),

$$\alpha = \sum_{k=0}^{\infty} \tilde{T}_k(\xi, \ldots, \xi). \tag{1.1.5}$$

It follows, from the uniqueness of the Ito–Wiener expansion, that (1.1.5) is the Ito–Wiener expansion α as a functional from ξ. Now the statement of the proposition trivially follows from the construction of $\tilde{T}_k,\ k \geq 0$.

Property 1.1.3. For the random variable $\alpha \in W^1$, the conditional expectation is $E(\alpha/\mathcal{F}_L) \in W^1$ and

$$D(E(\alpha/\mathcal{F}_L)) = PE(D\alpha/\mathcal{F}_L),$$

where $\mathcal{F}_L$ and P are from property 1.1.2.

Proof. Consider the Ito–Wiener expansion of α

$$\alpha = \sum_{k=0}^{\infty} Q_k(\xi, \ldots, \xi).$$

It follows, from the definition of the conditional mathematical expectation, that for every random variable $\beta \in L_2(\Omega, \mathcal{F}, \mathrm{P})$ which is measurable with respect to $\mathcal{F}_L$ the following equality holds

$$E\alpha\beta = EE(\alpha/\mathcal{F}_L) \cdot \beta. \tag{1.1.6}$$

Let the series

$$\beta = \sum_{k=0}^{\infty} T_k(\xi, \ldots, \xi)$$

be the Ito–Wiener expansion of β. Then, as in the proof of property 1.1.2,

$$\forall\, \varphi_1, \ldots, \varphi_k \in H_k: \quad T_k(\varphi_1, \ldots, \varphi_k) = T_k(P\varphi_1, \ldots, P\varphi_k), \ k \geq 1.$$

Therefore, from (1.1.6) and properties of Hilbert–Schmidt forms from ξ,

$$E(\alpha/\mathcal{F}_L) = \sum_{k=0}^{\infty} \tilde{Q}_k(\xi, \ldots, \xi),$$

where

$$\tilde{Q}_k(\ldots) = Q_k(P\,\cdot, \ldots, \cdot\, P), \ k \geq 0.$$

Now the statement of this property follows from the definition of the stochastic derivative.

Property 1.1.4. Let the function $f \in C^1(R)$ have a bounded derivative and $\alpha \in W^1$. Then $f(\alpha) \in W^1$ and

$$Df(\alpha) = f'(\alpha) D\alpha.$$

Proof. Operator D is closed. So it is sufficient to prove the statement only for the random variable $E(\alpha/\mathcal{F}_n)$, where $\mathcal{F}_n = \sigma(\{(\xi, e_t);\ 1 \leq t \leq n\})$. $\{e_t\,;\ t \geq 1\}$ is fixed orthonormal basis in H. Under a fixed $n \geq 1$, $E(\alpha/\mathcal{F}_n)$ depends only on $(\xi, e_1), \ldots, (\xi, e_n)$ and has a stochastic derivative. Hence

$$E(\alpha/\mathcal{F}_n) = f_n((\xi, e_1), \ldots, (\xi, e_n)),$$

where the function f_n belongs to $W_2^1(R^n, \mu)$, as in example 1.1.5. Consider the Ito–Wiener expansion of $E(\alpha/\mathcal{F}_n)$

$$E(\alpha/\mathcal{F}_n) = \sum_{k=0}^{\infty} Q_k(\xi, \ldots, \xi).$$

From property 1.1.2 it can be concluded that, for every $m \geq 1$, the sum

$$p_m = \sum_{k=0}^{m} Q_k(\xi, \ldots, \xi)$$

is the usual polynomial from $(\xi, e_1), \ldots, (\xi, e_n)$ and the sequence $\{p_m;\ m \geq 1\}$ converges in W^1 to $E(\alpha/\mathcal{F}_n)$. For every $m \geq 1$, $f(p_m) \in C^1(R^n)$ and the gradient of $f(p_m)$ is $f'(p_m)\nabla p_m$. So, from example 1.1.5, $f(p_m) \in W^1$ and

$$Df(p_m) = f'(p_m) Dp_m.$$

From this equality under $m \to \infty$,

$$Df(E(\alpha/\mathcal{F}_n)) = f'(E(\alpha/\mathcal{F}_n)) \cdot DE(\alpha/F_n).$$

The statement is proved.

Now consider some examples of stochastic derivatives.

Example 1.1.7. Consider the case, from example 1.1.2′, when ξ in $L_2([0;1])$ is constructed with the help of the Wiener process $\{w(t); t \in [0;1]\}$. It follows, from definition 1.1.1, that the random variable $w(t_0)$ under fixed t_0 is stochastically differentiable and

$$Dw(t_0) = \mathcal{X}_{[0;t_0]}.$$

Hence, if the function f satisfies the condition of property 1.1.4, the random variable $f(w(t_0))$ is stochastically differentiable and

$$Df(w(t_0)) = f'(w(t_0)) \cdot \mathcal{X}_{[0;t_0]}.$$

The random variable

$$\alpha = \int_0^1 \int_0^t a(t,s)dw(s)dw(t)$$

for $a \in L_2(\Delta_2(1))$, has the stochastic derivative

$$(D\alpha)(t) = \int_0^t a(t,s)\, dw(s) + \int_t^1 a(s,t)dw(s). \tag{1.1.7}$$

To prove this equality, note that α is a Hilbert–Schmidt form from ξ with the kernel

$$\tilde{a}(t,s) = \begin{cases} \frac{1}{2}a(t,s)\,, & s \le t\,, \\ \frac{1}{2}a(s,t)\,, & s > t\,. \end{cases}$$

Note that the equality (1.1.7) is the equality for two random elements in $L_2([0;1])$.

The second fundamental notion in stochastic calculus is the extended stochastic integral. Begin with the definition.

Definition 1.1.7. The conjugate operator $I = D^* : L_2(\Omega, \mathrm{P}, H) \to L_2(\Omega, \mathcal{F}, \mathrm{P})$ is called the extended stochastic integral.

It follows, from this definition and the properties of Hilbert–Schmidt forms from ξ, that every $Q_k(\xi, \ldots, \xi)$ $(Q_k \in H_k(H)\,, k \ge 0)$ belongs to the domain of definition $\mathcal{D}$ of I and

$$I(Q_k(\xi, \ldots, \xi)) = \Lambda \tilde{Q}_k(\xi, \overset{k+1}{\ldots}, \xi).$$

Here

$$\forall\, \varphi_1, \ldots, \varphi_{k+1} \in H: \quad \tilde{Q}_k(\varphi_1, \ldots, \varphi_{k+1}) = (Q_k(\varphi_1, \ldots, \varphi_k),\ \varphi_{k+1}), \tag{1.1.8}$$

where Λ is a symmetrization operator. Therefore, the definition of the extended stochastic integral can be formulated in terms of multiple Hilbert–Schmidt forms from ξ.

Definition 1.1.7′. The random element $x \in L_2(\Omega, \mathrm{P}, H)$ with the Ito–Wiener expansion

$$x = \sum_{k=0}^{\infty} Q_k(\xi, \ldots, \xi)$$

belongs to $\mathcal{D}$ if, and only if, the series

$$S = \sum_{k=0}^{\infty} \Lambda \tilde{Q}_k(\xi, \dots, \xi)$$

converges in $L_2(\Omega, \mathcal{F}, \mathrm{P})$, where $\tilde{Q}_k$ is organised from Q_k, as in (1.1.8), and $I(x) = S$.
Consider some examples and properties of I.

Example 1.1.8. $H = R^n, \xi = (\xi_1, \dots, \xi_n)$ as in example 1.1.1'. Consider the random vector $x = (p_1, \dots, p_n)$, where $p_1, \dots, p_n$ are polynomials from $\xi_1, \dots, \xi_n$. It follows, from definition 1.1.7', that $x \in \mathcal{D}$. To find $I(x)$ use definition 1.1.7. Consider the random variable $f(\xi_1, \dots, \xi_n)$, where $f \in C^1(R^n)$ and ∇f are bounded. Then

$$EI(x) \cdot f = E(x, Df) = E(x, \nabla f(\xi_1, \dots, \xi_n)).$$

From integration by parts, such a formula is obtained:

$$EI(x) \cdot f = E \sum_{j=1}^{n} \left(\xi_j p_j - \frac{\partial p_j}{\partial \xi_j} \right) \cdot f. \tag{1.1.9}$$

Hence

$$I(x) = \sum_{j=1}^{n} \xi_j p_j - \frac{\partial p_j}{\partial \xi_j}.$$

Denote by $W^1(H)$ the space of all stochastically differentiable random elements in H. The following statement is true.

Theorem 1.1.1. $W^1(H) \subset \mathcal{D}$ and for every $x \in W^1(H)$

$$E(I(x))^2 = E\|x\|^2 + E(Dx, (Dx)^*)_2.$$

Proof. For the case $x = Q_k(\xi, \dots, \xi)$, $Q_k \in H_k(H)$*and* $k \geq 0$, the statement of the theorem can be easily checked. To prove a general case use the Ito–Wiener expansion and an approximation procedure.

Example 1.1.9. As in example 1.1.7, consider the Wiener process $\{w(t) \ ; \ t \in [0; 1]\}$. Let the random process $\{x(t) \ ; \ t \in [0; 1]\}$ be adapted to the flow $\{\mathcal{F}_t = \sigma(\{w(s); \ s \leq t\}); \ t \in [0; 1]\}$ and

$$E \int_0^1 x^2(t) dt < \infty.$$

Consider x as a random element in $L_2([0; 1])$. Then $x \in \mathcal{D}$ and

$$I(x) = \int_0^1 x(t) dw(t),$$

where the integral on the right is the usual Ito integral. To prove this, consider the Ito–Wiener expansion of x

$$x(t) = \sum_{k=0}^{\infty} Q_k(\xi, \dots, \xi)(t).$$

It has been proved, in example 1.1.2′, that now $Q_k(\xi,\dots,\xi)(t)$, $k \geq 0$, are multiple Ito–Wiener integrals. The proof of property 1.1.2 of a stochastic derivative says that under t being fixed the kernels of $\{Q_k(t)\}$ are concentrated on $[0;t]^k$, $k \geq 1$. Therefore

$$x(t) = \sum_{k=0}^{\infty} \int_{\Delta_k(t)} q(t, \tau_1, \dots, \tau_k) dw(\tau_1) \dots dw(\tau_k).$$

Hence

$$I(Q_k(\xi,\dots,\xi)(\cdot)) = \int_{\Delta_{k+1}(1)} q(\tau_{k+1}, \tau_1, \cdot, \tau_k) dw(\tau_1) \dots dw(\tau_{k+1}).$$

So the series

$$S = \sum_{k=0}^{\infty} I\left(Q_k(\xi,\dots,\xi)(\cdot)\right) = \sum_{k=0}^{\infty} \int_{\Delta_{k+1}(1)} q(\tau_{k+1}, \tau_1, \dots, \tau_k) dw(\tau_1) \dots dw(\tau_{k+1})$$

converges and is equal to the Ito integral

$$S = \int_0^1 x(t) dw(t).$$

At the end of this chapter, other examples of stochastic integrals from random functions and their relations with the extended stochastic integral shall be considered.

Henceforth, the following fact will be useful.

Property 1.1.5.

$$\forall\, x \in \mathcal{D}: \quad EI(x) = 0.$$

Proof. From definition 1.1.7

$$\forall\, x \in \mathcal{D}: \quad EI(x) = EI(x) \cdot 1 = E(x, D1) = E(x, 0) = 0.$$

It follows, from the above consideration, that the stochastic integral has such a "bad" property. A starting point of their definition begins from the condition of a finite second moment. This assumption is not necessary for the Ito integral and is very strong when considering equations with the extended stochastic integral. In this chapter, operators D and I will be defined for larger classes of random variables and elements than W^1 and $\mathcal{D}$. To explain the idea of such an extension note that D and I are differential operators in a certain sense, as in examples 1.1.2′ and 1.1.7.

Hence D and I can be defined locally on some subset of Ω. The following section is devoted to the construction of a system of such subsets.

1.2 Smooth open sets

The system $\mathcal{E}$ of subsets for the localisation of D and I must have such a property. If random variables from W^1 or elements from D are equal on the set from $\mathcal{E}$, then the results of action D or I are also equal on the same set. It follows, from integration by parts of the formula, that this property holds if there is on set A a large class of random variables from W^1 which are equal to zero with its stochastic derivative for, almost all $\omega \notin A$.

Definition 1.2.1. The set $A \in \mathcal{F}$ is a smooth open set if there exists the random variable $\alpha \in W^1$ and the open set $G \subset R$ such that

$$A = \alpha^{-1}(G) \qquad (\text{mod P}).$$

Denote by $\mathcal{E}$ the system of all smooth open sets.

Consider examples.

Example 1.2.1. $H = R$, $\xi = \xi_1$ i.e. a Gaussian random variable with zero mean and variance 1. From section 1.1 it can now be concluded that $\alpha \in W^1$ if, and only if, $\alpha = f(\xi_1)$, where $f \in W_2^1(R, \mu)$ and μ is a distribution of ξ_1. It is known [3] that $W_2^1(R, \mu) \subset C(R)$. Therefore, for every open set $G \subset R$,

$$\alpha^{-1}(G) = \xi_1^{-1}(G_1),$$

where G_1 is open. Hence

$$\mathcal{E} = \{\xi_1^{-1}(G_1),\ G_1 \subset R - \text{ open }\}.$$

Example 1.2.2. Consider the situation of example 1.1.3. Check that $\mathcal{T} \subset \mathcal{E}$, where $\mathcal{T}$ is a class of all open subsets of B. Every open subset $A \subset B$ is a union of the countable set of balls because B is separable

$$A = \cup_{n=1}^{\infty} B(\theta_n, r_n),$$

$$B(\theta_n, r_n) = \{\theta \ : \|\theta - \theta_n\| < r_n\}\ , n \geq 1\,.$$

It has been proved, in the work of Sugita [4], that for every open ball, K, in B there exists a non-negative function f from $W_2^1(B, \mu)$ such that $B = \{f > 0\}$. The function f has a usual derivative along H.

Let $\{f_n\ ;\ n \geq 1\}$ be the sequence of such functions for $\{B(\theta_n, r_n)\ ;\ n \geq 1\}$. Then, for the certain sequence of positive numbers $\{a_n\ ; n \geq 1\}$, the function

$$g = \sum_{n=1}^{\infty} a_n f_n$$

is a random variable such that

$$A = g^{-1}((0;\infty))\,.$$

Hence $A \in \mathcal{E}$, which establishes the claim. When $\dim B = \infty$, $\mathcal{E}$ is larger than $\mathcal{T}$. To prove this it is sufficient to consider the random variable $\alpha = (\xi, \varphi)$, where $\varphi \in H \setminus B^*$. Then the function $\alpha\ : B \to R$ is only Borel measurable but not continuous. So there exists an open subset $G \subset R$, for which

$$\alpha^{-1}(G) \notin \mathcal{T}.$$

But $\alpha \in W^1$ and $\alpha^{-1}(G) \in \mathcal{E}$.

Consider the properties of smooth open sets.

For every set $A \in \mathcal{E}$ there are many random variables α and open subsets $G \subset R$ for which

$$A = \alpha^{-1}(G) \quad (\text{mod P}).$$

The following lemma gives α and G which will be useful in future.

Lemma 1.2.1. For every $A \in \mathcal{E}$ there exists a non-negative random variable $\alpha \in W^1$ such that
1) $\alpha \le 1 \quad (\text{mod P})$;
2) $\{\alpha > 0\} = A \quad (\text{mod P})$.

Proof. It follows, from the definition, that there exists $\beta \in W^1$ and open $G \subset R$ such that

$$A = \beta^{-1}(G) \quad (\text{mod P}).$$

There exists the non-negative function $f \in C^1(R)$ for which
1) $\forall\, x \in R \, : \, |\varphi'(x)| \le 1$,
2) $\{x \, : \, \varphi(x) > 0\} = G$.

Then the random variable $\alpha = \varphi(\beta) \in W^1$ and satisfies the conditions of the lemma.

In future every random variable which satisfies the conditions of lemma 1.2.1 will be called the canonical correspondent to set A.

Lemma 1.2.2. $\mathcal{E}$ is closed with respect to finite intersections and countable unions.

Proof. Consider $\{A_n \, ; \, n \ge 1\}$ i.e. the sequence of sets from $\mathcal{E}$. Let, for every $n \ge 1$, the random variable α_n canonically correspond to A_n. Then the random variable $\beta_N = \alpha_1 \cdot \ldots \cdot \alpha_N \in W^1$ and

$$\{\beta_N > 0\} = \bigcap_{n=1}^{N} A_n .$$

Hence $\cap_{n=1}^{N} A_n \in \mathcal{E}$.

Take, for every $n \ge 1$, $a_n > 0$ such that

$$\max\{a_n, a_n^2 \cdot E\|D\alpha_n\|^2\} < \frac{1}{4^n},$$

and consider

$$\gamma = \sum_{n=1}^{\infty} a_n \alpha_n .$$

The random variable $\gamma \in W^1$ and

$$\{\gamma > 0\} = \cup_{n=1}^{\infty} A_n .$$

Hence $\bigcup_{n=1}^{\infty} A_n \in \mathcal{E}$. Lemma is proved.

For future consideration a new definition is required.

Definition 1.2.2. For $B \in \mathcal{F}$, $A \in \mathcal{E}$ and $B \subset A$ the inclusion $B \subset A$ is called smooth if there is a random variable α canonically corresponding to A such that for some $c > 0$,

$$B \subset \{\alpha > c\}.$$

The smooth inclusion is denoted by

$$B \subset\subset A .$$

For $A \in \mathcal{F}$ $L_2(A)$ is a subspace of $L_2(\Omega, \mathcal{F}, P)$ which consists of all functions equal to zero outside A, $W_0^1(A)$ is a subspace of W^1 consisting of all functions for which its supports are smoothly included in A.

Lemma 1.2.3. For $A \in \mathcal{E}$, $W_0^1(A)$ is dense in $L_2(A)$.
Proof. It can be easily checked that the set of all bounded random variables from W^1 is complete in $L_2(\Omega, \mathcal{F}, P)$ using the Ito–Wiener expansion and property 1.1.4. Therefore, for every $\alpha \in L_2(A)$, there is a sequence $\{x_n \, ; \, n \geq 1\}$ of random variables for which

$$\forall\, n \geq 1 \ : \ x_n \in W^1 \cap L_\infty(\Omega, \mathcal{F}, P)\,,$$

$$E(x_n - \alpha)^2 \to 0\,,\ n \to \infty\,.$$

Hence, it is sufficient to prove that every x_n can be approximated on A in the square mean by elements of $W_0^1(A)$. Consider the random variable β canonically corresponding to A and for every $n \geq 1$ denote

$$\gamma_m = \varphi(m \cdot \beta)\,.$$

Here the non-decreasing function $\varphi \in C^1([0; +\infty))$ is such that

$$\forall\, s \leq \frac{1}{2} \ : \ \varphi(s) = 0\,,$$

$$\forall\, s \geq 1 \ : \ \varphi(s) = 1\,.$$

Then

$$\forall\, n \geq 1\,,\ m \geq 1 \ : \ \gamma_n \in W_0^1(A) \cap L_\infty(\Omega, \mathcal{F}, \mathrm{P}),$$

$$\gamma_n \cdot x_m \in W_0^1(A)$$

and

$$\forall\, m \geq 1 : \int_A (x_m - x_m \cdot \gamma_n)^2 d\mathrm{P} \to 0\,, n \to \infty\,.$$

Lemma is proved.

Now the localization of the stochastic derivative and the extended stochastic integral can be begun.

Theorem 1.2.1. Let for $A \in \mathcal{E}$, $\alpha_1\,, \alpha_2 \in W^1$, $x_1\,,\ x_2\ \in \mathcal{D}$

$$\mathcal{X}_A(\alpha_1 - \alpha_2) = 0\,,\ \mathcal{X}_A(x_1 - x_2) = 0 \quad (\text{mod P})\,.$$

Then

$$\mathcal{X}_A(D\alpha_1 - D\alpha_2) = 0\,, \mathcal{X}_A(I(x_1) - I(x_2)) = 0 \quad (\text{mod P})\,.$$

Proof. Begin with the statement for the extended stochastic integral. Note that for every bounded $\beta \in W_0^1(A)$,

$$\{\|D\beta\| > 0\} \subset\subset A\,.$$

Really, from the definition

$$\{|\beta| > 0\} \subset\subset A .$$

Hence, there exists a random variable γ canonically corresponding to A such, that

$$\{|\beta| > 0\} \subset \{\gamma > \frac{1}{2}\} .$$

Consider the random variable $\beta \cdot \varphi(2\gamma) \in W_0^1(A)$ where φ is from the proof of lemma 1.2.3. Then

$$D\beta = D(\beta \cdot \varphi(2\gamma)) = \varphi(2\gamma) \cdot D\beta + \beta \cdot 2\varphi'(2\gamma) \cdot D\gamma = \varphi(2\gamma) \cdot D\beta .$$

Therefore,

$$\{\|D\beta\| > 0\} \subset \{\gamma > \frac{1}{4}\} \subset\subset A .$$

Now for random elements x_1, x_2 from the theorem and bounded $\beta \in W_0^1(A)$.

$$0 = E(x_2 - x_1 , D\beta) = E(I(x_2) - I(x_1)) \cdot \beta .$$

It follows from the proof of the lemma 1.2.3 that the set of all bounded random variables from $W_0^1(A)$ is dense in $L_2(A)$. Hence,

$$\mathcal{X}_A(I(x_2) - I(x_1)) = 0 .$$

To prove the statement of the theorem for stochastic derivatives consider spaces $L_2(A, H)$ and $W_0^1(A, H)$ which are similar to $L_2(A)$ and $W_0^1(A)$ but consist of random elements in H. It can be proved as in lemma 1.2.3 that bounded random elements form $W_0^1(A, H)$ are dense in $L_2(A, H)$, and that for every bounded $\beta \in W_0^1(A, H)$

$$\{\|D\beta\|_2 > 0\} \subset\subset A .$$

From the statement of the theorem for the extended stochastic integral which was proved above we have

$$\forall\, x \in W_0^1(A, H) \ : \ \{|I(x)| > 0\} \subset\subset A .$$

It is sufficient now to use the definition of the extended stochastic integral as adjoint operator to a stochastic derivative to finish the proof of the theorem.

It must be mentioned that for stochastically differentiable random elements x_1 and x_2 condition $A \in \mathcal{E}$ is not necessary and the statement of the theorem remains true in the case when $A \in \mathcal{F}$. This fact was proved by E. Pardoux and D. Nualart [5]. Let us consider their theorem.

Theorem 1.2.2. Let $x_1, x_2 \in W^1(H), A \in \mathcal{F}$ and

$$\mathcal{X}_A(x_2 - x_1) = 0 \quad (\text{mod P}) .$$

Then

$$\mathcal{X}_A(I(x_2) - I(x_1)) = 0 \quad (\text{mod P}) .$$

Proof. Let $\varphi \in C^2(R)$ be a non-negative function and

$$\varphi(0) = 1, \ \text{supp}\, \varphi \subset [-1; 1].$$

Then the random variable $\alpha_\varepsilon = \varphi(\frac{1}{\varepsilon}\cdot\|x_2 - x_1\|^2) \in W^1$ and

$$D\alpha_\varepsilon = \frac{2}{\varepsilon}\varphi'(\frac{1}{\varepsilon}\cdot\|x_2 - x_1\|^2)\sum_{k=1}^{\infty}\sum_{j=1}^{\infty}[(D(x_2 - x_1, e_j), e_k)(e_j, x_2 - x_1)]\cdot e_k\,, \quad (1.2.1)$$

where $\{e_k\,,\ k \geq 1\}$ - an orthonormal basis in H. To prove (1.2.1) it is sufficient to check it in the case, when x_1, x_2 are polynomials from the finite number of random variables $\{(\xi, e_k)\,,\ k \geq 1\}$ and to use the property 1.1.1 of the operator D. Now for the bounded random variable $\beta \in W^1$

$$E(I(x_2) - I(x_1))\cdot\alpha_\varepsilon\beta = EI(x_2 - x_1)\cdot\alpha_\varepsilon\beta =$$

$$= E(x_2 - x_1, D(\alpha_\varepsilon\beta)) = E(x_2 - x_1, \beta D\alpha_\varepsilon) + E(x_2 - x_1, \alpha_\varepsilon D\beta) =$$

$$= E\beta\sum_{k,j=1}^{\infty}(x_2 - x_1, e_k)(x_2 - x_1, e_j)\,(D(x_2 - x_1, e_j), e_k)\,\times$$

$$\times\frac{2}{\varepsilon}\varphi'(\frac{1}{\varepsilon}\|x_2 - x_1\|^2) + E\alpha_\varepsilon(x_2 - x_1, D\beta)\,.$$

Hence, from the Lebesgue dominated convergence theorem under $\varepsilon \to 0$

$$EI(x_2 - x_1)\cdot\mathcal{X}_{\{x_2 = x_1\}}\beta = E(x_2 - x_1, D\beta)\cdot\mathcal{X}_{\{x_2 = x_1\}} = 0\,.$$

So,

$$(I(x_2) - I(x_1))\cdot\mathcal{X}_{\{x_2 = x_1\}} = 0\,,$$

and

$$I(x_2) - I(x_1)\cdot\mathcal{X}_A = 0\,.$$

The theorem is proved.

An analogous statement can be proved for stochastic derivatives by using the function $\psi \in C^2(R)$ with supp $\psi \subset [-1;1]$, $\psi(0) = 0$, $\psi'(0) = 1$.

1.3 The localization of the extended stochastic integral and the stochastic derivative

From theorems 1.2.1 and 1.2.2 we can conclude that operators D and I can be extended on random variables, an elements which have not a finite second moment. To develop not only smooth random elements we can use localization with the help of theorem 1.2.1.

Definition 1.3.1. The random element $x \in H$ stochastically integrable if there exist sequences $\{A_n; n \geq 1\}$, $\{x_n; n \geq 1\}$ of smooth open sets and random elements from D such, that

1) $A_n \subset A_{n+1}$, $n \geq 1$, $\cup_{n=1}^{\infty} A_n = \Omega$ (mod P),

2) $\forall\, n \geq 1:\ x_n\cdot\mathcal{X}_{A_n} = x\cdot\mathcal{X}_{A_n}$ (mod P).

The random variable α such that

$$\forall n \geq 1:\ I(x_n)\cdot\mathcal{X}_{A_n} = \alpha \ \ (\text{mod P})$$

is called the extended stochastic integral of x and denoted by $I(x)$.

Definition 1.3.2. The random variable α is stochastically differentiable if there exist sequences $\{A_n;\ n \geq 1\}$, $\{\alpha_n; n \geq 1\}$ of smooth open sets and random variables from W^1 such that
1) $A_n \subset A_{n+1},\ n \geq 1,\ \cup_{n=1}^{\infty} A_n = \Omega$ (mod P),
2) $\forall\, n \geq 1:\ \alpha_n \cdot \mathcal{X}_{A_n} = \alpha \cdot \mathcal{X}_{A_n}$ (mod P).
The random element z in H such that

$$\forall\, n \geq 1:\ D\alpha_n \cdot \mathcal{X}_{A_n} = z \cdot \mathcal{X}_{A_n} \quad (\text{mod P})$$

is called the stochastic derivative of α and denoted by $D\alpha$.

The correctness of definition 1.3.1 and 1.3.2 follows from the theorem 1.2.1. Note, that stochastically differentiable random elements in real separable Hilbert space can be defined similar to the definition 1.3.2. In future the set of all stochastically differentiable random elements in real separable Hilbert space $\mathcal{H}$ is denoted by $\tilde{W}^1(\mathcal{H})$ (in the case $\mathcal{H} = R$ is simply $\tilde{W}^1$). The set of all stochastically integrable random elements in H is denoted by $\tilde{\mathcal{D}}$. The definition and designation for higher stochastic derivative are analogous.

Let us consider some examples.

Example 1.3.1. The case $H = R$. Consider a random variable $\alpha = f(\xi_1)$, where $F \in C^1(R)$. Then $\alpha \in \tilde{W}^1$. To prove this it is enough to use such sequences of smooth open sets and random variables from W^1:

$$\forall\, n \geq 1:\ A_n = (-n; n),$$

$$\alpha_n = f_n(\xi),\ f_n(t) = \begin{cases} f(n), & t > n, \\ f(t), & -n \leq t \leq n, \\ f(-n), & t < -n. \end{cases}$$

In the case of example 1.1.3 the random variable α which corresponds to the function f such that $f'_H \in C(B)$ is stochastically differentiable. To prove this, let us consider the sequence of open sets $\{G_n; n \geq 1\}$ in B such that

$$\forall\, n \geq 1:\ G_n \subset G_{n+1},$$

$$\sup\{\theta \in G_n:\ |f(\theta)|, \|f'_H(\theta)\|\} \leq n, \cup_{n=1}^{\infty} G_n = B.$$

Denote by g_1 the function which was built in the example 1.2.2 for open set G_1. Let $\Delta_n = G_n \setminus \bar{G}_{n-1}, n \geq 2$, where $\bar{G}_{n-1}$ – the closure of G_{n-1}. Functions $\{h_n; n \geq 1\}$ are constructed in such a way for $\{\Delta_n;\ n \geq 1\}$ as g_1 for G_1. Denote by $g_n = g_1 + h_2 + \cdots + h_n,\ n \geq 1$. Consider $h_{nm} = \varphi(m \cdot g_n), m \geq 1$, where φ is a non-decreasing function from $C^1(R)$ such that $\varphi(0) = 0, \varphi([1; +\infty)) = \{1\}$. Then sequences

$$G'_n = \{g_n > \frac{1}{n}\},\ \alpha_n = f \cdot h_{nn}$$

have the following properties

$$\forall\, n \geq 1:\ G'_n \in \mathcal{E}, \alpha_n \in W^1,$$

$$\mathcal{X}_{G'_n} \cdot f = \alpha_n \mathcal{X}_{G'_n} \quad (\text{mod } \nu),$$

$$G'_n \subset G'_{n+1}, n \geq 1.$$

The measure ν is Gaussian. So, for every $n \geq 1$

$$\nu(\partial G_n) = 0.$$

Hence

$$\cup_{n=1}^{\infty} G'_n = B \pmod{\nu}.$$

Therefore $\alpha \in W^1$. Our claim is proved.

Example 1.3.1 shows, that the definition 1.3.2 envelops random variables, which are smooth from intuitive senses, but have not a finite second moment. The following example is devoted to the reverse situation.

Example 1.3.2. Let $\alpha \in W^1$ and $c \in R$ be such that

$$P\{\alpha = c\} = 0.$$

Then the characteristic function $\mathcal{X}$ of the set $\{\alpha > c\}$ belongs to $\tilde{W}^1$. Really, consider the non-decreasing function $\psi \in C^1(R)$ such that

$$\psi((-\infty; 0]) = \{0\}, \psi([1; +\infty)) = \{1\}.$$

Then

$$\forall\, n \geq 1 : \ \psi(n \cdot (\alpha - c)) \in W^1,$$

and on the set $\{\alpha < c\} \cup \{\alpha > c + \frac{1}{n}\}$

$$\psi(n \cdot (\alpha - c)) = \mathcal{X}.$$

Therefore $\mathcal{X} \in \tilde{W}^1$ and

$$D\mathcal{X} = 0 \pmod{\mathrm{P}}.$$

Note that $\mathcal{X} \in L_2(\Omega, \mathcal{F}, \mathrm{P}) \cap \tilde{W}^1$, but there exist situations in which $\mathcal{X} \bar{\in} W^1$. For example, let $H = R$. Now as in the example 1.1.4, all random variables from W^1 are absolutely continuous functions from ξ. So $\mathcal{X}_{(0;+\infty)} \bar{\in} W^1$, but

$$\mathcal{X}_{(0;+\infty)} = \mathcal{X}_{\{\xi > 0\}} \in \tilde{W}^1.$$

The following lemma will be used in future.

Lemma 1.3.1. Let the random variable $\alpha \in \tilde{W}^1$, $G \subset R$ be open. Then $\alpha^{-1}(G) \in \mathcal{E}$.

Proof. Consider sequences $\{\alpha_n;\ n \geq 1\}$ $\{A_n;\ n \geq 1\}$ of random variables from W^1 and smooth open sets, which correspond to α in the definition 1.3.2. Let the non-negative function $\varphi \in C^1(R)$ has a bounded derivative and be such that

$$\{\varphi > 0\} = G.$$

It can be easily checked that the random variable $\beta = \varphi(\alpha)$ satisfies conditions of the definition 1.3.2 with sequences $\{\beta_n = \varphi(\alpha_n);\ n \geq 1\}$ and $\{A_n;\ n \geq 1\}$. Consider for every $n \geq 1$ the sequence of random variables $\{\gamma_{nm};\ m \geq 1\}$, which was built in the proof of the lemma 1.2.3 for the set A_n. Then

$$\forall\, n, m \geq 1 : \ \beta_n \cdot \gamma_{nm} \in W_0^1(A_n),$$

$$\beta_n \cdot \gamma_{nm} = \beta \cdot \gamma_{nm}, \ 0 \le \beta \cdot \gamma_{nm} \le \beta\gamma_{nm+1} \le \beta \quad (\text{mod P}).$$

Now

$$\{\beta > 0\} = \cup_{n,m-1}\{\beta\gamma_{nm} > 0\} \in \mathcal{E}$$

from the lemma 1.2.2. Lemma is proved.

Now we can formulate the theorem about a stochastic derivative of superposition without technical conditions.

Theorem 1.3.1. Let $F \in C^1(R^n), \alpha_1, \ldots, \alpha_n \in \tilde{W}^1$. Then $F(\alpha_1, \ldots, \alpha_n) \in \tilde{W}^1$ and

$$DF(\alpha_1, \ldots, \alpha_n) = \sum_{j=1}^{N} \frac{\partial F}{\partial \alpha_j} D\alpha_j.$$

Proof. Consider the sequence of functions $\{F_k;\ k \ge 1\}$ such that
1) $\forall\ k \ge 1: \ F_k \in C^1(R^n)$,

$$\sup_{R^n} |F_k| < +\infty, \ \sup_{R^n} \|\nabla F_k\| < +\infty,$$

2) $\forall\ k \ge 1: \ \text{Int}\ \{F_k = F\} \subset \text{Int}\{F_{k+1} = F\}$,

$$\cup_{k=1}^{\infty}\{F_k = F\} = R^n.$$

For every $j = 1, \ldots, n$ let $\{A_{jk};\ k \ge 1\}$, $\{\alpha_{jk};\ k \ge 1\}$ be localizing sequences from smooth open sets and stochastically differentiable random variables from the definition 1.3.2 for the random variable α_j. Then for every $k \ge 1$ the set

$$B_k = \cap_{j=1}^{n} A_{jk} \cap \{(\alpha_1, \ldots, \alpha_n) \in Int\{F_k = F\}\}$$

is smooth open. This follows from lemmas 1.2.2 and 1.3.1. For the random variable $F_k(\alpha_{1k}, \ldots, \alpha_{nk}) \in W^1$

$$DF_k(\alpha_{1k}, \ldots, \alpha_{nk}) = \sum_{j=1}^{n} \frac{\partial F_k}{\partial \alpha_{jk}} D\alpha_{jk}, \ k \ge 1.$$

Hence $F(\alpha_1, \ldots, \alpha_n) \in \tilde{W}^1$ and

$$DF(\alpha_1, \ldots, \alpha_n) = \sum_{j=1}^{n} \frac{\partial F}{\partial \alpha_j} D\alpha_j.$$

Theorem is proved.

It follows from analogous localizing arguments that the following theorem is true.

Theorem 1.3.2. $\tilde{W}^1(H) \subset \tilde{\mathcal{D}}$.

In the following two sections we shall consider a localizing extended stochastic integral and a stochastic derivative in the situations of the examples 1.1.2 and 1.1.3.

1.4 Stochastic integrals with respect to Gaussian random measures

Let us consider the situation of the example 1.1.2. Now $H = L_2(\mathcal{X}, \mathcal{A}, \mu)$ and ξ is constructed with the help of the Gaussian random measure γ on $\mathcal{A}$ such that

$$\forall\, A, B \in \mathcal{A}: \ E\gamma(A) = 0, \ E\gamma(A)\cdot\gamma(B) = \mu(A \cap B).$$

In this case random elements in H are random functions [6] on $\mathcal{X}$ and the definition of I is the definition of certain stochastic integral with respect to γ. There are several construction of the stochastic integral with respect to γ. In this section we shall consider some of them and construct the class of simple random functions for which the definition of a stochastic integral is natural. The relation between the extended stochastic integral and Stratonovich integral will be considered in chapter 3. Note that some aspects of the construction of the stochastic integral with respect to Gaussian random measures are considered in many papers (see for example [7 – 10]).Our aim in this section is to unify different definitions of the stochastic integral.

The known definitions of the integral

$$\int_{-} x(\theta,\omega)\gamma(d\theta)$$

can be divided into two groups. The Stratonovich integral and the integrals constructed in the papers [11 – 13] belong to the first group. Here and in other papers, close to them, the construction of the integral does not take into account the nature of the probable dependence between the random function x and the measure γ. The Ito integral, the extended stochastic integral, and the Pardoux-Protter integral [14] which will be discussed in future, belong to the second group. Now we consider the class of random functions for which the definition of the stochastic integral is very natural.

We suppose that the measure μ has no atoms and all the random functions considered satisfy the condition

$$\mathrm{P}\left\{\int_{-} x^2(\theta,\omega)\mu(d\theta) < \infty\right\} = 1.$$

Let $A \in \mathcal{A}$ be such that $\mu(A) > 0$, $\mu(\mathcal{X} \setminus A) > 0$. We denote by $\mathcal{F}_A$ the σ-algebra generated by the random quantities $\{\gamma(C):\ C \in \mathcal{A},\ C \subset \mathcal{A}\}$.

Definition 1.4.1. If x is a measurable random function with respect to the σ-algebra $\mathcal{F}_{--A}$,then by integral

$$\int_A x(\theta,\omega)\gamma(d\theta)$$

we mean the sum

$$\sum_{n=1}^{-} \int_A \varphi_n(\theta)\gamma(d\theta)\cdot\int_A \varphi_n(\theta)x(\theta,\omega)\mu(d\theta). \tag{1.4.1}$$

Here $\{\varphi_n; n \geq 1\}$ is an orthonormal basis in $L_2(A, \mu)$.

The following lemma ensures that definition 1.4.1 is well posed.

Lemma 1.4.1. The series (1.4.1) converges in probability and its sum does not depend on the choice of a basis.

Proof. For each $m \in N$ we define the random function x_m as follows:

$$x_m(\theta,\omega) = \begin{cases} x(\theta,\omega), & \int_A x^2(\theta,\omega)\mu(d\theta) < m, \\ 0, & \int_A x^2(\theta,\omega)\mu(d\theta) \geq m. \end{cases}$$

The series (1.4.1) for x_m is mean-square convergent. In fact, for all $1 \leq n_1 \leq n_2$ we have

$$E\left(\sum_{n=n_1}^{n=n_2} \int_A \varphi_n(\theta)\gamma(d\theta) \cdot \int_A \varphi_n(\theta)x_m(\theta,\omega)\mu(d\theta)\right)^2 =$$

$$= \sum_{n=n_1}^{n=n_2} E\left(\int_A \varphi_n(\theta)x_m(\theta,\omega)\mu(d\theta)\right)^2 .$$

For any other orthonormal basis $\{\psi_n;\ n \geq 1\}$ in $L_2(A,\mu)$ we have

$$E\left(\sum_{n=1}^{\infty}\left(\int_A \varphi_n(\theta)\gamma(d\theta) \cdot \int_A \varphi_n(\theta)x_m(\theta,\omega)\mu(d\theta) - \right.\right.$$

$$\left.\left. - \int_A \psi_n(\theta)\gamma(d\theta)\right) \cdot \int_A \psi_n(\theta)x_m(\theta,\omega)\mu(d\theta)\right)^2 =$$

$$= E\sum_{n=1}^{\infty}\left(\int_A \varphi_n(\theta)x_m(\theta,\omega)\mu(d\theta)\right)^2 + E\sum_{n=1}^{\infty}\left(\int_A \psi_n(\theta)x_m(\theta,\omega)\mu(d\theta))\right)^2 -$$

$$-2\sum_{n=1}^{\infty}\sum_{k=1}^{\infty}\int_A \varphi_n(\theta)\psi_k(\theta)\mu(d\theta)E\left[\int_A \varphi_n(\theta)x_m(\theta,\omega)\mu(d\theta) \cdot \int_A \psi_k(\theta)x_m(\theta,\omega)\mu(d\theta)\right] =$$

$$= 2E\int_A x_m^2(\theta,\omega)\mu(d\theta) - 2\sum_{n=1}^{\infty} E\left(\int_A \varphi_n(\theta)x_m(\theta,\omega)\mu(d\theta)\right)^2 =$$

$$= 2E\int_A x_m^2(\theta,\omega)\mu(d\theta) - 2E\int_A x_m^2(\theta,\omega)\mu(d\theta) = 0.$$

Consequently, the assertion of the lemma holds for x_m. Since the right-hand sides of the series (1.4.1) for x_m and x are the same on the set $\{x_m = x\}$, the assertion of the lemma holds also for the function x in view of the equality

$$\cup_{m=1}^{\infty}\{x_m = x\} = \Omega \quad (\text{mod P}).$$

The lemma is proved. The integral which we have introduced possesses the following property.

Lemma 1.4.2. For every function x satisfying the conditions of definition 1.4.1 the following inequality holds for all $c > 0$, $\varepsilon > 0$

$$\mathrm{P}\left\{\left|\int_A x(\theta,\omega)\gamma(d\theta)\right| > \varepsilon\right\} \leq \frac{c}{\varepsilon^2} + \mathrm{P}\left\{\left|\int_A x^2(\theta,\omega)\mu(d\theta)\right| > c\right\}.$$

Proof. For any $\varepsilon > 0$

$$\mathrm{P}\left\{|\int_A x(\theta,\omega)\gamma(d\theta)| > \varepsilon\right\} = \int_\Omega \mathrm{P}\left(\left\{|\int_A x(\theta,\omega)\gamma(d\theta)| > \varepsilon\right\} / \mathcal{F}_{--A}\right) d\mathrm{P}.$$

The random quantity

$$\mathrm{P}\left(\left\{|\int_A x(\theta,\omega)\gamma(d\theta)| > \varepsilon\right\} / \mathcal{F}_{--A}\right)$$

admits the estimate

$$\mathrm{P}\left(\left\{|\int_A x(\theta,\omega)\gamma(d\theta)| > \varepsilon\right\} / \mathcal{F}_{--A}\right) \leq \min\left(\frac{1}{\varepsilon^2} \cdot \int_A x^2(\theta,\omega)\mu(d\theta)\ ,1\right).$$

Consequently, for every $c > 0$

$$\mathrm{P}\left(\left\{|\int_A x(\theta,\omega)\gamma(d\theta)| > \varepsilon\right\} / \mathcal{F}_{--A}\right) \leq \frac{1}{\varepsilon^2} c \cdot \mathcal{X}_{-\int_A x^2(\theta,\omega)\mu(d\theta) - c-} + \mathcal{X}_{-\int_A x^2(\theta,\omega)\mu(d\theta) > c-}.$$

Hence the required inequality follows.

Definition 1.4.2. A random function x is said to be simple if there exists a partition of $\mathcal{X}$ into a finite number of disjoint measurable subsets $A_1, \ldots, A_n$ such that for every $i = 1, \ldots, n$ the function $\mathcal{X}_{A_i} \cdot x$ is measurable with respect to σ-algebra $\mathcal{F}_{--A_i}$.

Definition 1.4.3. By the integral of the simple random function x with respect to the measure γ and the set $B \in \mathcal{A}$ we mean the sum

$$\int_B x(\theta,\omega)\gamma(d\theta) := \sum_{i=1}^n \int_{B-A_i} x(\theta,\omega)\gamma(d\theta). \tag{1.4.2}$$

The collection of sets $A_1, \ldots, A_n$ in the definition 1.4.2 may be chosen in many ways; however, as the next lemma shows, the sum on the right-hand side of (1.4.2) does not depend on this choice.

Lemma 1.4.3. Let for the $A_1, \ldots, A_n$ and $C_1, \ldots, C_n$ be two different collections of sets satisfying the definition 1.4.2 for the simple random function x. Then for arbitrary $B \in \mathcal{A}$

$$\sum_{i=1}^n \int_{B-A_i} x(\theta,\omega)\gamma(d\theta) = \sum_{j=1}^n \int_{B-C_j} x(\theta,\omega)\gamma(d\theta).$$

Proof. We set for all $i = 1, \ldots, n$, $j = 1, \ldots, m$, $D_{ij} = B \cap C_j \cap A_i$. For those i, j for which $D_{ij} \neq \emptyset \pmod{\mu}$, the $\{\varphi_{kij};\ k \geq 1\}$ form an orthonormal basis in $L_2(D_{ij}, \mu)$. Then by lemma 1.4.1

$$\sum_{i=1}^n \int_{B-A_i} x(\theta,\omega)\gamma(d\theta) = \sum_{i=1}^n \sum_{j=1}^m \sum_{k=1}^{-} \int_{D_{ij}} \varphi_{kij}(\theta)\gamma(d\theta) \cdot \int_{D_{ij}} \varphi_{kij}(\theta)\mu(d\theta) =$$

$$= \sum_{j=1}^n \int_{B\cap C_j} x(\theta,\omega)\gamma(d\theta).$$

The lemma is proved.

It can be easily checked in similar fashion that the above integral has the additional property that is, for every simple function x and any disjoint sets $A, B \in \mathcal{A}$ the following equality holds:

$$\int_{A\cup B} x(\theta,\omega)\gamma(d\theta) = \int_A x(\theta,\omega)\gamma(d\theta) + \int_B x(\theta,\omega)\gamma(d\theta).$$

Moreover, we have the following properties.

Property 1.4.1. Let x_1, x_2 be simple random functions, $\alpha, \beta \in R$. Then $\alpha x_1 + \beta x_2$ is a simple random function and for all $A \in \mathcal{A}$

$$\int_A (\alpha x_1 + \beta x_2)(\theta,\omega)\gamma(d\theta) = \alpha \int_A x_1(\theta,\omega)\gamma(d\theta) + \beta \int_A x_2(\theta,\omega)\gamma(d\theta).$$

Property 1.4.2. Let $\{x_n;\ n \ge 1\}$ be a sequence of simple functions with respect to a fixed partition $A_1, \ldots, A_m$ of the set $\mathcal{X}$. Suppose that for some random function x

$$\int_{\mathcal{X}} (x_n - x)^2(\theta,\omega)\mu(d\theta) \xrightarrow{P} 0,\ n \to \infty.$$

Then x is a simple random function with respect to the partition $A_1, \ldots, A_m$ and for every $A \in \mathcal{A}$

$$\int_A x_n(\theta,\omega)\gamma(d\theta) \xrightarrow{P} \int_A x(\theta,\omega)\gamma(d\theta),\ n \to \infty.$$

Property 1.4.1 is obvious while property 1.4.2 follows from definitions 1.4.2 and 1.4.3 and the lemma 1.4.2.

Remark. The definition 1.4.3 of the integral of a simple function x with respect to measure γ can be treated as follows. For some orthonormal basis $\{\varphi_n;\ n \ge 1\}$ in $L_2(\mathcal{X}, \mathcal{A}, \mu)$ we have for all $A \in \mathcal{A}$

$$\int_A x(\theta,\omega)\gamma(d\theta) = \sum_{i=1}^{\infty} \int_A \varphi_n(\theta)\gamma(d\theta) \cdot \int_A \varphi_n(\theta)x(\theta,\omega)\mu(d\theta).$$

However the basis $\{\varphi_n;\ n \ge 1\}$ cannot be chosen arbitrary as following example shows.

Example 1.4.1. Let $\mathcal{X} = A_1 \cup A_2$, $A_1 \cap A_2 = \emptyset$, $A_1, A_2 \in \mathcal{A}$, $\mu(A_1) > 0$, $\mu(A_2) > 0$. We denote by $\{\varphi_n;\ n \ge 1\}$, $\{\psi_n;\ n \ge 1\}$ orthonormal bases in $L_2(A_1, \mu)$ and $L_2(A_2, \mu)$ respectively. Consider the random function.

$$x(\theta,\omega) = \sum_{n=1}^{\infty} \frac{1}{n} \cdot \frac{1}{\sqrt{2}} \left[\varphi_n(\theta) \cdot \int_{A_2} \psi_n(\theta)\gamma(d\theta) + \psi_n(\theta) \cdot \int_{A_1} \varphi_n(\theta)\gamma(d\theta)\right].$$

The series on the right-hand side defines x only as a random element in $L_2(\mathcal{X}, \mu)$; however, according to [6], there exists a version of x that is measurable in required way. The function x is simple with respect to the partition A_1, A_2.

Let $\{e_k;\ k \geq 1\}$ be an orthonormal basis in $L_2(\mathcal{X}, \mu)$; of such kind $\{\frac{1}{\sqrt{2}}(\varphi_n \pm \psi_n);\ n \geq 1\}$. Then the series

$$\sum_{n=1}^{\infty} \int_{\mathcal{X}} e_k(\theta)\gamma(d\theta) \cdot \int_{\mathcal{X}} e_k(\theta)x(\theta,\omega)\mu(d\theta)$$

has as its nonzero terms a sequence of random quantities of the form

$$\frac{1}{n} \cdot \frac{1}{2}\Big(\int_{A_1} \varphi_n(\theta)\gamma(d\theta) \pm \int_{A_2} \psi_n(\theta)\gamma(d\theta)\Big) \cdot$$

$$\Big(\int_{A_2} \psi_n(\theta)\gamma(d\theta) \pm \int_{A_1} \varphi_n(\theta)\gamma(d\theta)\Big),\ n \geq 1$$

and may fail to converge in probability depending on the order of the terms.

Theorem 1.4.1. The collection of simple random functions is dense in

$$L = L_2(\Omega \times \mathcal{X}, \mathcal{F} \times \mathcal{A}, \mathrm{P} \times \mu).$$

Every simple random function x in L occurs in $\tilde{\mathcal{D}}$ and

$$I(x) = \int_{\mathcal{X}} x(\theta,\omega)\gamma(d\theta).$$

For every y in $\mathcal{D}$ there exists a sequence of simple functions $\{x_n;\ n \geq 1\}$ in L such that

$$E\int_{\mathcal{X}} (x_n(\theta,\omega) - y(\theta,\omega))^2\mu(d\theta) \to 0,\ n \to \infty.$$

$$E\Big(\int_{\mathcal{X}} x_n(\theta,\omega)\gamma(d\theta) - I(y)\Big)^2 \to 0,\ n \to \infty.$$

Proof. Let $z \in L$. Since the measure μ is finite and non-atomic and the σ-algebra $\mathcal{A}$ is countably generated, it follows [15] that there exists a sequence of partitions of $\mathcal{X}$ into disjoint measurable subsets $\{A_{nk};\ k = 1, \ldots, n;\ n \geq 1\}$ such that

$$\max_{k=1,\ldots,n} \mu(A_n k) \to 0,\ n \to \infty.$$

For convenience we denote by $\mathcal{F}_{nk}$ the σ-algebra $\mathcal{F}_{\bar{A}_{nk}}$, $k = 1, \ldots, n;\ n \geq 1$, and by P_{nk} the projection in $L_2(\mathcal{X}, \mu)$ onto the subspace $L_2(A_{nk}, \mu)$. We set

$$\forall\, n \geq 1:\ z_n = \sum_{k=1}^{n} E(P_{nk}z/\ \mathcal{F}_{nk}). \qquad (1.4.3)$$

Using Ito–Wiener expansion of z it can be verified that

$$E\Big(\int_{\mathcal{X}} (z_n(\theta,\omega) - z(\theta,\omega))^2\mu(d\theta)\Big) \to 0,\ n \to \infty.$$

Suppose now that x is a simple random function in L with respect to the partition $A_1, \ldots, A_n$. We consider Ito–Wiener expansion of x

$$x(\theta,\omega) = \sum_{k=0}^{\infty} \int_{\mathcal{X}^k} f_k(\theta, \theta_1, \ldots, \theta_k)\gamma(d\theta_1)\ldots\gamma(d\theta_k).$$

The non-random function f_k, $k \geq 1$, have the following property (the proof of the property 1.1.2)

$$\forall\, i = 1,\dots,n,\ k \geq 1, \theta \in A_i, \theta_1,\dots,\theta_k \in \mathcal{X},$$
$$f_k(\theta,\theta_1,\dots,\theta_k) = f_k(\theta,\theta_1,\dots,\theta_k)\cdot \mathcal{X}_{\bar{A}_i}(\theta_1)\cdot\ldots\cdot \mathcal{X}_{\bar{A}_i}(\theta_k).$$

Therefore for each $k \geq 1$ the random function

$$\int_{\mathcal{X}^k} f_k(\theta,\theta_1,\dots,\theta_k)\gamma(d\theta_1)\cdot\ldots\cdot\gamma(d\theta_k)$$

is simple with respect to the partition $A_1,\dots,A_n$.

By definition 1.4.2

$$\int_{\mathcal{X}}\Big(\int_{\mathcal{X}_k} f_k(\theta,\theta_1,\dots,\theta_k)\gamma(d\theta_1)\cdot\ldots\cdot\gamma(d\theta_k)\Big)\gamma(d\theta) =$$

$$= \sum_{i=1}^{n}\sum_{j=1}^{\infty}\int_{\mathcal{X}}\varphi_{ij}(\theta)\gamma(d\theta)\cdot\int_{\mathcal{X}}\int_{\mathcal{X}^k} f_k(\theta,\theta_1,\dots,\theta_k)\gamma(d\theta_1)\cdot\ldots\cdot\gamma(d\theta_k)\varphi_{ij}(\theta)\mu(d\theta) =$$

$$= \sum_{i=1}^{n}\sum_{j=1}^{\infty}\int_{\mathcal{X}}\varphi_{ij}(\theta)\gamma(d\theta)\cdot\int_{\mathcal{X}^k}\Big(\int_{\mathcal{X}} f_k(\theta,\theta_1,\dots,\theta_k)\varphi_{ij}(\theta)\mu(d\theta)\Big)\gamma(d\theta_1)\cdot\ldots\cdot\gamma(d\theta_k) =$$

$$= \int_{\mathcal{X}^{k+1}} \Lambda f_k(\theta,\theta_1,\dots,\theta_{k+1})\gamma(d\theta_1)\cdot\ldots\cdot\gamma(d\theta_{k+1}),$$

where $\{\varphi_{ij};\ j \geq 1\}$ is an orthonormal basis in $L_2(A_i,\ \mu)$. In now follows from property 1.4.2 of our integral of simple functions and because the operator I is closed

$$\int_{\mathcal{X}} x(\theta,\omega)\gamma(d\theta) = \sum_{k=0}^{\infty}\int_{\mathcal{X}^{k+1}}\Lambda f_k(\theta_1,\dots,\theta_{k+1})\gamma(d\theta_1)\dots\gamma(d\theta_{k+1}) = I(x).$$

The proof of the final assertion of the theorem is carried out in similar fashion using the construction (1.4.3) and Ito–Wiener expansion. The theorem is proved.

To illustrate this theorem let us consider an example of construction of the stochastic integral in the particular case: $\mathcal{X} = [0;1]$, $\gamma(d\theta) = dw(\theta)$, $\{w(\theta);\ \theta \in [0;1]\}$ - Wiener process.

Example 1.4.2. Two-sided stochastic integral of Pardoux-Protter (sf. [14]). Consider the class $\mathcal{M}$ of random functions on $[0;1]$ of such kind

$$\forall:\ x(\theta) = f(u(\theta), v(\theta), \theta),\ \theta \in [0;1].$$

Here $f:\ R^2 \times [0;1] \to R$ is a smooth function, u and v are solutions of stochastic differential equations with initial conditions at the point 0 and 1 respectively. The integral of a random function from $\mathcal{M}$ is defined [14] as the limit in probability of the following sums, which are analogous to integral sums:

$$S = \sum_{k=0}^{n-1} f(\theta_k, u(\theta_k), v(\theta_{k+1}))(w(\theta_{k+1}) - w(\theta_k)) \tag{1.4.4}$$

under the condition that the gauge of the partition tends to 0. Note that the sum (1.4.4) is the integral from simple function

$$x_n = \sum_{k=0}^{n-1} f(\theta_k, u(\theta_k), v(\theta_{k+1}))\mathcal{X}_{[\theta_k;\theta_{k+1})}.$$

Hence, by the theorem 1.4.1

$$S = I(x_n).$$

It is proved in [14] that for every $x \in \mathcal{M} \cap L$ the limit of sums (1.4.4) exists in square-mean. Since the operator I is closed, the two-sided stochastic integral coincides with the extended stochastic integral for all $x \in \mathcal{M} \cap L$. Moreover it can be proved, that if equations for u and v have smooth coefficients then $x \in \tilde{W}^1(L_2([0;1]))$.

To prove this it is sufficient to use such facts. Firstly, the random function u (and v) belongs to $\tilde{W}^1(L_2([0;1]))$ (see [16]). Secondly, under the smooth coefficients

$$\forall\, L > 0\; \exists\, K > 0:$$

$$\{\omega : \max_{[0;1]} |w| < L\} \subset \{\omega : \max_{[0;1]} |u| < K\}.$$

Hence we can use the localization on the sequence of smooth open sets

$$A_n = \{\omega : \max_{[0;1]} |w| < n\},\ n \geq 1.$$

So, for $x \in \mathcal{M}$ with smooth coefficients in equations for u and v, the two-sided stochastic integral exists and coincides with the extended stochastic integral (the detail proof can be founded in the article of the author [17])

1.5 Sobolev spaces on infinitely-dimensional domain

The procedure of the localization of operators D and I in the situation of the example 1.1.3 is closely related to the construction of Sobolev spaces. In this section we consider definition of Sobolev space on infinitely-dimensional domain and their certain simple properties.

Let the space B be Hilbert space. Consider an open subset $G \subset B$.

Definition 1.5.1. Sobolev space $W^1(G, \mu)$ of first order on the set G with respect to the measure μ is the completion of the set of the functions

$$L = \{f :\ f \in C^1(G),\ \int_G (f^2 + \|f'_H\|^2) d\mu < +\infty\}.$$

with respect to the norm, which is generated by the inner product

$$\forall\, f, g \in L:\ (f, g)_1 := \int_G (f \cdot g + (f'_H, g'_H)) d\mu.$$

The theorem 1.2.1 ensures that this definition is well-posed. The generalized derivative is defined for all elements of $W^1(G, \mu)$ in a usual way. In the case $G = B$ the derivative can be obtained with the help of the Ito–Wiener expansion. Consider the case of arbitrary open G. There are some new aspects in this situation. For example the equality $f \equiv \text{const}$ do not follows from the equality $Df \equiv 0$. To check this it is sufficient to consider $B = R$, $G = (0;1) \cup (2;3)$. This difficulty can be overcome if we suppose that the set G is coherent.

Theorem 1.5.1. Let the set G be coherent in B, and for the function $f \in W^1(G, \mu)$

$$Df = 0 \quad (\text{mod } \mu).$$

Then there exists $a \in R$ such, that

$$f = a \quad (\text{mod } \mu).$$

Proof. The set G is a linear coherent in B. To prove the note that for any $x \in G$ the set A_x of all points from G, which may be continuously connected with x is open. And $G \setminus A_x$ is open. Hence it is enough to prove the theorem in the case, when G – is an open ball in B. Let S be a covariance operator of the measure μ. Denote by $\{e_k;\ k \geq 1\}$ own orthonormal basis of S and by $\{\lambda_k;\ k \geq 1\}$ corresponding own values. Note that any open ball $B(x,r)$ in B can be presented as such a union

$$B(x,r) = \cup_{n=n_0}^{\infty} B(P_n x, r - |x - P_n x|),$$

where P_n is a projector onto subspace which is generated by $\{e_1, \ldots, e_n\}$. Number n_0 is such that $P_n x_0 \in B(x,r)$. Since

$$\forall\, n_1,\ n_2 \geq n_0 :\ \mu(B(P_{n_1}x, r - |x - P_{n_1}x|) \cap B(P_{n_2}x, r - |x - P_{n_2}x|)) > 0,$$

then it is enough to prove the theorem for the open ball which has a center in the union $\cup_{n=1}^{\infty} P_n(B)$. From now consider $B(x_0, r)$ where $x_0 \in \cup_{n=1}^{\infty} P_n(B)$. The proof can be divided into two parts. Firstly, present $B(x_0, r)$ as

$$B(x_0, r) = \cup_{n=1}^{\infty} Z_n,$$

where

$$\forall\, n \geq 1 :\ Z_n = Q_{n1} \times Q_{n2} \times \cdots Q_{nm} \times \cdots.$$

Here $\{Q_{nm};\ n, m \geq 1\}$ – open balls in finite-dimensional spaces. Under this

$$\forall\, n_1, n_2 \geq 1 :\ \mu(Z_{n_1} \cap Z_{n_2}) > 0.$$

Secondly, prove our claim for the function f on the set Z_n, $n \geq 1$.

For any $n \geq 1$ the set Z_n can be constructed in such a way. Denote by $\varepsilon_n = \frac{1}{2^{n+1}}\mu(B(x_0, r))$. Take such $r_1 \in (0; r)$ that

$$\mu(B(x_0, r) \setminus B(x_0, r_1)) < \frac{1}{2}\varepsilon_n.$$

After that take k_1 such that

$$(I - P_{k_1})x_0 = 0,$$

$$\bar{\mu}_{k_1}(B_{\bar{k}_1}(0; \frac{1}{2}\delta)) > 1 - \frac{1}{4}\varepsilon_n.$$

Here $\bar{\mu}_{k_1}$ – projection of measure μ on the closed linear envelope of $\{e_{k_1+1}, \ldots\}$ [18], $\delta = r - r_1$, $B_{\bar{k}_1}(0,\delta)$ – the open ball with the center 0 and the radius δ in the closed linear envelope of $\{e_{k_1+1}, \ldots\}$. The possibility of such a choice follows from the condition on x_0 and Chebyshev inequality

$$\bar{\mu}_{\bar{k}}(B_{\bar{k}}(0;\delta)) > 1 - \frac{\sum_{i=k+1}^{\infty} \lambda_i}{\delta} \to 1, \ k \to \infty.$$

Let

$$Q_{n1} = P_{k_1}(B(x_0, r_1)).$$

Take $k_2 > k_1$ such that

$$\bar{\mu}_{k_2}(B_{k_2}(0; \frac{\delta}{2})) > 1 - \frac{1}{8}\varepsilon_n,$$

and

$$Q_{n2} = P_{k_1 k_2}(B_{\bar{k}_1}(0,\delta)),$$

where $P_{k_1 k_2}$ – the projector on the linear envelope $\{e_{k_1+1}, \ldots, e_{k_2}\}$. If we continue this procedure, then we obtain the sequence of open balls $\{Q_{nm};\ m \geq 1\}$ in finite-dimensional subspaces such, that

$$\forall\, i \geq 2 : \ Q_{ni} = P_{k_{i-1} k_i}(B_{\bar{k}_{i-1}}(0, \frac{1}{2^{i-1}}\delta)),$$

$$\bar{\mu}_{k_i k_i}(B_{\bar{k}_i}(0, \frac{1}{2^{i-1}}\delta)) > 1 - \frac{1}{2^{i+1}}\varepsilon_n.$$

Let

$$Z_n = Q_{n1} \times Q_{n2} \times \cdots Q_{nm} \times \cdots.$$

Then Z_n is a Borel subset of $B(x_0, r)$. Note that

$$\mu(Z_n) \geq \mu(B(x_0, r_1))(1 - \frac{1}{2}\varepsilon_n) \cdot (1 - \frac{1}{4}\varepsilon_n) \cdot \ldots \geq$$

$$\geq \mu(B(x_0, r_1)) \cdot (1 - \varepsilon_n) \geq \mu(B(x_0, r)) - \frac{3}{2}\varepsilon_n.$$

Hence

$$\mu(B(x_0, r) \setminus Z_n) < \frac{1}{2^n}\mu(B(x_0, r)).$$

Therefore the sequence $\{Z_n;\ n \geq 1\}$ satisfies our conditions.

Let now $f \in W^1(B(x_0, r), \mu)$ and

$$Df = 0 \quad (\mathrm{mod}\ \mu).$$

Then there exists the sequence of functions $\{f_n;\ n \geq 1\}$ such that

1) $f_n \in C^1(B(x_0, r))$ and f_n, f'_n are bounded on $B(x_0, r)$, $n \geq 1$,

2) $\int_{B(x_0,r)}((f_n - f)^2 + \|Df_n\|^2)d\mu \to 0, n \to \infty.$

For any $l, m, n \geq 1$ the function

$$f_{lmn}(P_{k_m}x) = \int_{Q_{nm+1}\times\cdots} f_l(P_{k_m})x + (I - P_{k_m})x)d\bar{\mu}_{k_m}$$

has the continued bounded derivative on the set $Q_{n1} \times Q_{n2} \times \cdots Q_{nm}$. Moreover

$$\int_{Q_{n1}\times Q_{n2}\times\cdots Q_{nm}} (f_{lmn}(P_{k_m}x) - f_{mn}(P_{k_m}x))^2 d\mu_{k_m} \to 0, \ l \to \infty,$$

and the sequence of all derivatives converges to 0 in the square-mean. Here

$$f_{mn}(P_{k_m}x) = \int_{Q_{nm+1}\times\cdots} f(P_{k_m}x + (I - P_{k_m})x)d\bar{\mu}_{k_m}.$$

Hence

$$f_{mn} \in W^1(Q_{n1} \times Q_{n2} \times \cdots Q_{nm}, \ \mu_{k_m})$$

and the generalized derivative of f_{mn} is equal to 0 almost sure up to the measure μ_{k_m} on $Q_{n1} \times Q_{n2} \times \cdots Q_{nm}$. Note, that $Q_{n1} \times Q_{n2} \times \cdots Q_{nm}$ is an open bounded set in the finite-dimensional space and μ_{k_m} - a non-degenerate Gaussian measure on this space. Hence $W^1(Q_{n1} \times Q_{n2} \times \cdots Q_{nm}, \ \mu_{k_m})$ coincides with the Sobolev space of first order, which is built on the set $Q_{n1} \times Q_{n2} \times \cdots Q_{nm}$ with respect to the Lebesgue measure λ. Therefore for any $\varphi \in W^1(Q_{n1} \times Q_{n2} \times \cdots Q_{nm}, \ \mu_{k_m})$ the Poincare inequality [19] is true

$$\int_{Q_{n1}\times Q_{n2}\times\cdots Q_{nm}} \varphi^2 d\lambda \leq K \left(\int_{Q_{n1}\times Q_{n2}\times\cdots Q_{nm}} \|D\varphi\|^2 d\lambda + \frac{1}{\lambda(Q_{n1} \times Q_{n2} \times \cdots Q_{nm})} \times \right.$$

$$\left. \times \left(\int_{Q_{n1}\times Q_{n2}\times\cdots Q_{nm}} \varphi d\lambda \right)^2 \right),$$

where the constant K is not depending from φ.

Consequently,

$$\forall \, n, m \geq 1 : \exists \, a_{nm} \in R :$$

$$f_{nm} = a_{nm} \quad (\text{mod } \mu_{k_m})$$

on $Q_{n1} \times Q_{n2} \times \cdots Q_{nm}$. It follows now from the Levy theorem [20] that f is a constant μ-almost sure on Z_n. So f is a constant μ-almost sure on $B(x_0, r)$. The theorem is proved.

There is an interesting consequence of this theorem.

Consequence 1.5.1. Let G be an open coherent set in B, $f \in W^1(G, \mu)$. Then the support of the measure μf^{-1} on R is a coherent set.

Proof. Suppose that our claim is not true. Then there exist values $c_1 < c_2$ such that

$$G = \{f < c_1\} \cup \{f > c_2\} \quad (\text{mod } \mu),$$

$$\mu(\{f < c_1\}) > 0, \ \mu(\{f > c_2\}) > 0.$$

Consider the function φ on R, which has a continuous bounded derivatives and such, that

$$\forall \, t \leq c_1 : \ \varphi(t) = 0, \varphi'(t) = 0,$$

$$\forall\, t \geq c_2 : \ \varphi(t) = 1, \varphi'(t) = 0.$$

It can be easily checked, that $\varphi(f) \in W^1(G,\mu)$ and $D\varphi(f) = \varphi'(f) \cdot Df = 0 \pmod{\mu}$. Then from theorem 1.5.1, for the certain $a \in R$ $\varphi(f) = a \pmod{\mu}$ on G. But $\varphi(f) = 0$ on $\{f < c_1\}$, and $\varphi(f) = 1$ on $\{f < c_2\}$. This contradiction established our proposition.

Therefore functions from $W^1(G,\mu)$ has up to the measure μ some properties of continuous functions.

Chapter 2

Random maps in Hilbert space

2.1 The definition of a random map

The aim of the chapter 1 was to prepare instruments for solving different types of equations including the extended stochastic integral and its generalizations. For example, in chapter 5, the following equation will be considered

$$x(t) + \int_0^1 K(t,s)x(s)ds + \int_0^1 L(t,s)x(s)dw(s) = y(t), \; t \in [0,1],$$

where the known and unknown random processes, y and x, may not be adapted to the flow of σ-algebras of the Wiener process w, and the second integral in the left part is the extended stochastic integral. Before studying such an equation consider its left part as an original object. For any $\varphi \in L_2([0;1])$, under suitable conditions on K and L, the random process

$$F(\varphi)(t) = \varphi(t) + \int_0^1 K(t,s)\varphi(s)ds + \int_0^1 L(t,s)\varphi(s)dw(s), t \in [0;1]$$

can be regarded as the random element in $L_2([0;1])$. In this case the map F has the following property: F is linear and, if the kernels are bounded, is continue in the square-mean. Note that, in the general case, the map F cannot be generated with the help of the set of bounded linear operators $\{\tilde{F}(\omega), \omega \in \Omega\}$ by the equality

$$\forall \varphi \in L_2([0;1]) : \; F(\varphi)(\omega) = \tilde{F}(\omega)(\varphi) \; (\text{mod P}).$$

In the case $L(t,s) = \mathcal{X}_{[0;t]}(s)$ this has been noted by A.V. Skorokhod [21]. More general examples are to be considered. Moreover, the map F is measurable with respect to the Wiener process w. So for the investigation of F, and to determine the action of F on random elements (random functions) in $L_2([0;1])$, the stochastic calculus developed in chapter 1 can be used.

Consider the general definition.

Let X be a separable real Hilbert space with the inner product $[\cdot,\cdot]$ and norm $|\cdot|$.

Definition 2.1.1. The random map F in space X is a map under which, for every element $x \in X$, $F(x)$ is a random element in X such that

1) $\forall\, x \in X:\ E|F(x)|^2 < +\infty$,
2) $E|F(x) - F(x_0)|^2 \to 0, |x - x_0| \to 0$.

The notion of a random map is a generalization of the notion of a strong random operator which has been introduced by A.V. Skorokhod [21]. Recall this definition for further considerations.

Definition 2.1.2. The strong random operator A in X is a map under which, for every element $x \in X$, $A(x)$ is a random element in X such that
1) $\forall\, \alpha, \beta \in R, x, y \in X$:
$A(\alpha x + \beta y) = \alpha A(x) + \beta A(y)$ (mod P),
2) $A(x) \xrightarrow{P} A(x_0), |x - x_0| \to 0$.

It follows, from the theorem about a closed graph, that the strong random operator A satisfying the condition

$$\forall\, x \in X : E|x|^2 < +\infty$$

is a random map. An important example of a strong random operator is a Gaussian strong random operator.

Definition 2.1.3. The strong random operator A is a Gaussian random operator if, for every $x \in X$, $A(x)$ has a Gaussian distribution in X.

Such operators will be considered in detail in this chapter.

2.2 Gaussian strong random operators

In future the abbreviation GSRO will be used instead of the words "Gaussian strong random operator". The large set of GSRO can be obtained with the help of the integration with respect to the Gaussian random X-valued measures.

Definition 2.2.1. W is a Gaussian random X-valued measure with independent values on disjoint sets and a structure measure λ on the Boreal σ-algebra $\mathcal{B}$ of subsets of $[0; 1]$ if
1) for any disjoint $\Delta_1, \Delta_2 \in \mathcal{B}$ $W(\Delta_1)$ and $W(\Delta_2)$ are independent Gaussian random elements in X,
2) $\forall\, \Delta_1, \Delta_2 \in \mathcal{B}$:
$EW(\Delta_1) = 0, E[W(\Delta_1), W(\Delta_2)] = \lambda(\Delta_1 \cap \Delta_2)$.

Note that the mathematical expectations in 2) exist because a Gaussian random element in the Banach space has all the moments of the norm [18]. For any function $f \in L_2([0; 1], \lambda)$, the stochastic integral

$$\int_0^1 f(t) W(dt)$$

is defined in the usual way [22]. Consider the GSRO

$$\forall\, x \in X:\ A(x) := \int_0^1 (Ux)(t) W(dt), \tag{2.2.1}$$

where U is a bounded linear operator with dense $U(X)$ in $L_2([0; 1], \lambda)$. The knowledge of the structure of W and U gives information about the properties of the

GSRO A. Before discussing corresponding results, consider the conditions under which the GSRO A has the representation (2.2.1).

Define the mathematical expectation of the GSRO A such that

$$\forall\, x \in X:\ (EA)(x) := EA(x).$$

It is easy to check that MA is a bounded linear operator in space X. Since the goal is to study stochastic properties of GSRO, consider now the GSRO A for which $MA = 0$. The distributions of all random elements $\{A(x); x \in X\}$ are uniquely defined by such moment functions

$\forall\, x, y, u, v \in X:$

$\beta(x, u; y, v) := E[A(x), u] \cdot [A(y), v].$

So the conditions for GSRO A to have the representation (2.2.1) can be formulated in terms of β. Let

$$U_0 = \sqrt{MA^*A},$$

where

$$\forall x, y \in X: [(MA^*A)(x), y] = M[A(x), A(y)].$$

Then

$$\forall x, u, y, v \in X:\ \beta(x, u; y, v) = [\Phi_{u,v} U_0 x, U_0 y],$$

where, for any $u, v \in X$, $\Phi_{u,v}$ is a bounded linear operator on the subspace $\overline{U_0(X)} = X_0$. Operators $\Phi_{u,v}$ are linearly dependent on u and v. The following theorem is true.

Theorem 2.2.1. GSRO A has the representation (2.2.1) if, and only if, $\{\Phi_{u,v};\ u, v \in X\}$ is a family of commuting self-adjoint operators.

Proof. *Necessity.* Let GSRO A have the representation (2.2.1). Denote by $S(\Delta)$ the covariance operator of the distribution of $W(\Delta)$. $S(\Delta)$ is a nuclear self-adjoint non-negative operator and

$$\forall\, \Delta \in \mathcal{B}:\ \ \mathrm{tr}\, S(\Delta) = \lambda(\Delta).$$

Moreover, for any sequence of disjoint sets $\{\Delta_n;\ n \geq 1\} \subset \mathcal{B}$,

$$S(\cup_{n=1}^{\infty} \Delta_n) = \sum_{n=1}^{\infty} S(\Delta_n),$$

where the series converges in the nuclear norm. There exists [23] the derivative $\frac{dS}{d\lambda}$. Denote it by R. For any $t \in [0; 1]$ $R(t)$ is a nuclear, self-adjoint, non-negative operator with $\mathrm{tr}\, R = 1$ and

$$\forall u, v \in X, \Delta \in \mathcal{B}:$$

$$[S(\Delta)u, v] = \int_{\Delta} [R(t)u, v]\lambda(dt).$$

Consider in $L_2([0; 1], \lambda)$ such a family of commuting self-adjoint operators $\{\tilde{\Phi}_{u,v}, u, v \in X\}$

$$\forall f \in L_2([0; 1], \lambda):$$

$$(\tilde{\Phi}_{u,v} f)(t) = [R(t)u, v] \cdot f(t).$$

Then the moment function of GSRO A has such a representation:

$$\beta(x,u;y,v) = \int_0^1 (\tilde{\Phi}_{u,v}Ux)(t) \cdot (Uy)(t)\ \lambda(dt).$$

Note that

$$\forall\, x \in X:\ |U_0x|^2 = E[Ax, Ax] = \int_0^1 (Ux)^2(t)\lambda(dt).$$

Since $\overline{U(X)} = L_2([0;1])$, then there exists an isometric map $J : X_0 \to L_2([0;1], \lambda)$ such that $\forall\, x \in X:\ Ux = JU_0x$. Note that

$$\forall u, v \in X : \Phi_{u,v} = J^{-1}\tilde{\Phi}_{u,v}J.$$

Hence, the family $\{\Phi_{u,v};\ u, v \in X\}$ satisfies the conditions of the theorem.

Sufficiency. Let the family $\{\Phi_{u,v};\ u, v \in X\}$ satisfy the conditions of the theorem. Denote by $\{e_i;\ i \geq 1\}$ the orthonormal basis in X. It follows from [23] that, on σ-algebra which is generated by cylindrical sets in R^{N^2}, there exists the decomposition of unity E such that

$$\forall\, i, j \geq 1:\ \Phi_{e_i,e_j} = \int_{R^{N^2}} \alpha_{ij} E(d\alpha),$$

where α_{ij} is the ij-element of $\alpha \in R^{N^2}$. It follows from inequalties

$$\forall\, i, j \geq 1:\ \|\Phi_{e_i,e_j}\| \leq |e_i| \cdot |e_j| = 1,$$

that

$$E(\prod\nolimits_{i,j=1}^{\infty}[-1,2)) = I,$$

where I is an identity operator in X_0. There exists (for example, [15]) a mutually measurable bijection

$$g:\ \prod_{i,j=1}^{\infty}[-1;2) \to [0;1].$$

Then $G = Eq^{-1}$ is a decomposition of the unity of B and

$$\forall i, j = 1 : \Phi_{e_i,e_j} = \int_0^1 \varphi_{e_i,e_j}(t)G(dt),$$

$$\varphi_{e_i,e_j} = g_{ij}^{-1}.$$

Operators $\Phi_{u,v}$ continuously depend on $u, v \in X$. So, for every $u, v \in X$ there exists a Boreal function $\varphi_{u,v}$ such that

$$\Phi_{u,v} = \int_0^1 \varphi_{u,v}(t)G(t),$$

$$\|\Phi_{u,v}\| = \|\varphi_{u,v}\|_\infty.$$

Here $\|\varphi_{u,v}\|_\infty$ is an essential upper bound of $|\varphi_{u,v}|$ with respect to measure G. Since $\Phi_{u,v} = \Phi_{v,u}$, then

$$\forall\, u, v \in X:\ \varphi_{u,v} = \varphi_{v,u} \pmod{G}.$$

To prove that $\Phi_{u,v} = \Phi_{v,u}$, consider such equalities

$$\forall\, x, y, u, v \in X :$$

$$[\Phi_{u,v} U_0 x, U_0 y] = \beta(x, u; y, v) =$$
$$= \beta(y, v; x, u) = [\Phi_{v,u} U_0 y, U_0 x] = [\Phi_{v,u} U_0 x, U_0 y].$$

Moreover

$$\forall\, u \in X :\ \Phi_{u,u} \geq 0.$$

Hence

$$\forall\, u \in X :\ \varphi_{u,u} \geq 0 \quad (\text{mod } G).$$

The important property of the family $\{\varphi_{u,v};\ u, v \in X\}$ follows from the equality

$$f, g \in X_0 :\ \sum_{i=1}^{\infty} [\Phi_{e_i,e_i} f, g] = [f, g].$$

If $f = g = G(\Delta)h, \Delta \in \mathcal{B}$, where h is an element of the maximal spectral type for measure G, then

$$\int_{\Delta} \sum_{i=1}^{\infty} \varphi_{e_i,e_i}(t)[G(dt)h, h] = [G(\Delta)h, h].$$

So

$$\sum_{i=1}^{\infty} \varphi_{e_i,e_i} = 1 \quad (\text{mod } G).$$

Let

$$\mathcal{M} = \{u \in X :\ [u, e_i] \in Q,\ i \geq 1, \exists i_0 = i_0(u) :\ \forall\, i \geq i_0 :\ [u, e_i] = 0\}.$$

It follows, from the properties of functions $\{\varphi_{u,v};\ u, v \in X\}$, that there exists a Boreal set $\Delta_0 \in \mathcal{B}$ such that $G(\Delta_0) = I$ and, for any $t \in \Delta_0$, there exists a bounded linear operator $R(t)$ on $\mathcal{M}$ for which
1) $\forall\, u, v \in \mathcal{M} :\ [R(t)u, v] = \varphi_{u,v}(t)$,
2) $\forall\, u, v \in \mathcal{M} :\ [R(t)u, v] = [R(t)v, u]$,
3) $\forall\, u \in \mathcal{M} :\ [R(t)u, u] \geq 0$
4) $\sum_{i=1}^{\infty} [R(t)e_i, e_i] = 1$.

Continue $R(t)$ on all X_0 as a bounded operator. $R(t)$ is a self-adjoint, non-negative nuclear operator with tr $R(t) = 1$. Take T to be a self-adjoint, non-negative, nuclear operator with tr $T = 1$. Define, for $t \bar{\in} \Delta_0$, $R(t) = T$. Then

$$\forall\, u, v \in X :\ \varphi_{u,v} = (R(\cdot)u, v) \quad (\text{mod } G).$$

Hence

$$\forall x, y, u, v \in X : \beta(x, u; y, v) = \int_0^1 [R(t)u, v] \cdot [G(dt)U_0 x, U_0 y].$$

Consider an operator $B = \int_0^1 tG(dt)$. There exists a self-adjoint, bounded operator C in X_0, with a simple spectrum, such that B is a function from C. To prove this, use the spectral theorem in the form of multiplication operators [24]. There exists the measurable space $(X, \mathcal{A}, \mu)$ with a probability measure μ, the unitary operator

$K : X \to L_2(\mathcal{X}, \mathcal{A}, \mu)$ and the function $\psi \in L_\infty(\mathcal{X}, \mathcal{A}, \mu)$ such that KBK^{-1} is a multiplication on ψ. The space $\mathcal{X}$ can be chosen as a countable union of copies of R and $\mathcal{A}$ as a Boreal σ-algebra of $\mathcal{X}$. There exists a mutually measurable bijection $g : [0; 1] \to \mathcal{X}$. Define an isometric map $Q : L_2(\mathcal{X}, \mathcal{A}, \mu) \to L_2([0; 1], \mathcal{B}, \mu g)$ in such a way:

$$\forall \varphi \in L_2(\mathcal{X}, \mathcal{A}, \mu) : \ Q\varphi = \varphi(g).$$

Then $QKBK^{-1}Q^{-1}$ is a multiplication on $\psi(g)$. So $QKBK^{-1}Q^{-1} = \psi(g)(D)$, where D is a multiplication on a free variable. The operator D has a simple spectrum. Therefore, it can be supposed that measure G has a simple spectrum. Then there exists a probability measure, λ on $\mathcal{B}$, and an isometry, $J : X_0 \to L_2([0; 1], \lambda)$, such that

$$\forall\, f \in L_2([0; 1], \lambda) :$$

$$J \int_0^1 tG(dt) J^{-1}(f)(s) = s \cdot f(s) \quad (\text{mod } \lambda).$$

Denote $U = JU_0$. Then

$$\forall\, x, y, u, v \in X :$$

$$\beta(x, u; y, v) = \int_0^1 [R(t)u, v] \cdot (Ux)(t) \cdot (Uy)(t) \lambda(dt),$$

$$E|Ax|^2 = |U_0 x|^2 = \int_0^1 (Ux)^2(t) \lambda(dt).$$

Since $\overline{UX} = L_2([0; 1], \lambda)$, then for every $f \in L_2([0; 1], \lambda)$ there exists a sequence $\{x_n;\ n \geq 1\} \subset X$ such that

$$\lim_{n\to\infty} Ux_n = f.$$

Define

$$\Phi f = \lim_{n\to\infty} Ax_n.$$

The value Φf does not depend on the sequence $\{x_n;\ n \geq 1\}$. For every $f \in L_2([0; 1], \lambda)$, Φf is a Gaussian random element in X with zero mean and

$$\forall f, g \in L_2([0; 1], \lambda) \quad \forall u, v \in X :$$

$$E[\Phi f, u] \cdot [\Phi g, v] = \int_0^1 [R(t)u, v] f(t) g(t) \lambda(dt).$$

So the correspondence

$$\mathcal{B} \ni \Delta \longrightarrow W(\Delta) := \Phi(\mathcal{X}_\Delta)$$

is a Gaussian X-valued measure with independent values on disjoint sets and the structure measure λ. Moreover

$$\forall f \in L_2([0; 1], \lambda) :$$

$$\Phi f = \int_0^1 f(t) W(dt).$$

Hence

$$\forall\, x \in X : \ Ax = \int_0^1 (Ux)(t) W(dt).$$

The theorem is proved.

To display an example of the usage of expansion (2.2.1) definitions from [21] are needed.

Definition 2.2.2. The strong random operator B in X is bounded with the probability 1 (or simply bounded), if there exists a family of linear, bounded, non-random operators $\{B(\omega), \omega \in \Omega\}$ in X such that

$$\forall\, x \in X : (Bx)(\omega) = B(\omega)x \quad (\text{mod P}).$$

Definition 2.2.3. A weak random operator B in X is the map, which corresponds for every $x, y \in X$ to the random variable $[Bx, y]$, with properties
1) $\forall\, x_1, x_2, y_1, y_2 \in X, \alpha_1, \alpha_2, \beta_1, \beta_2 \in R$:

$$[B(\alpha_1 x_1 + \alpha_2 x_2), \beta_1 y_1 + \beta_2 y_2] = \sum_{i,j=1}^{2} \alpha_i \beta_j [Bx_i, y_j],$$

2) $\forall\, x_0, y_0 \in X$:

$$P - \lim_{x \to x_0, y \to y_0} [Bx, y] = [Bx_0, y_0].$$

Definition 2.2.4. The weak random operator B^* is an adjoint to the weak random operator B if

$$\forall\, x, y \in X : [B^* x, y] = [By, x].$$

Denote the sets of bounded, strong and weak operators as $L_b(\Omega, X)$, $L_s(\Omega, X)$, $L_w(\Omega, X)$ respectively. Inclusions

$$L_b(\Omega, X) \subset L_s(\Omega, X) \subset L_w(\Omega, X)$$

are obvious. The conditions necessary for $B \in L_w(\Omega, X)$ to belong to $L_s(\Omega, X)$ easily follow from the Minlos-Sazonov theorem. The conditions for which GSRO A belongs to $L_b(\Omega, X)$ will be of interest now. The inclusion $A^* \in L_s(\Omega, X)$ is weaker than $A \in L_b(\Omega, X)$. To check this consider the following example.

Example 2.2.1. Let $\{e_k;\ k \geq 1\}$ be an orthonormal basis in X and $\{\xi_k;\ k \geq 1\}$ be a sequence of independent, identically distributed Gaussian random variables with zero mean and variance 1. Consider such GSRO in X

$$\forall x \in X : \ Ax := \sum_{k=1}^{\infty} \xi_k \cdot [e_k, x] \cdot e_k.$$

It can be easily checked that $A = A^* \in L_s(\Omega, X)$. But $A \bar{\in} L_b(\Omega, X)$. This follows from the equality

$$P\{\sup_{k \geq 1} |Ae_k| = \sup_{k \geq 1} |\xi_k| = +\infty\} = 1.$$

If the GSRO has the representation of theorem 2.2.1 and its adjoint belongs to $L_s(\Omega, X)$, then the random measure W must be of a very special kind.

Theorem 2.2.2. Let GSRO A, with zero mean, have the representation from theorem 2.2.1. Suppose that the operator U is a bijection between a certain subspace

X_1 of X and $L_2([a;b],\lambda)$, where λ is a structure measure for W. If $A^* \in L_s(\Omega, X)$, then the measure W is discrete on $[a;b]$.

Proof. Suppose that the claim is false. Then there exists a Borel subset $E \subset [a;b]$ such that $\lambda(E) > 0$ and every point from E has λ- measure zero. It follows from the proof of theorem 2.2.1, that the moment function β of GSRO A can be written in the form:

$$\forall\, x, y, u, v \in X:$$

$$\beta(x,u;y,v) = \int_0^1 (Ux)(t)(Uy)(t)\cdot[R(t)u,v]\lambda(dt).$$

Since

$$\forall\, t \in [0;1]: \ \text{tr}\ R(t) = 1,$$

then there exist a Borel subset $\Delta \subset E$ and an element $e \in X$ such that

$$\lambda(\Delta) > 0,$$

$$\forall\, t \in \Delta: [R(t)e, e] > 1.$$

Then, for any $x \in U^{-1}(L_2(\Delta,\lambda)) \cap X_1$,

$$\beta(x,e;x,e) = \int_\Delta (Ux)^2(t)\cdot[R(t)e,e]\lambda(dt) \geq$$

$$\geq \int_\Delta (Ux)^2(t)\lambda(dt) \geq \|U_1\|^{-1}\cdot|x|^2,$$

where U_1 is a construction of U on X_1. It follows, from the choice of E and Δ, that

$$\dim U^{-1}(L_2(\Delta,\lambda)) \cap X_1 = \dim L_2(\Delta,\lambda) = \infty.$$

So there are no nuclear operators with such a bilinear form as

$$B(x,y) = \beta(x,e;y,e), x, y \in X.$$

But, from the condition $A^* \in L_s(\Omega, X)$,

$$B(x,y) = M[A^*e,x]\cdot[A^*e,y],\ x,y \in X.$$

Consequently, the form B corresponds to the covariance operator of A^*e, which is nuclear. This contradiction establishes the claim. Theorem is proved.

To consider the conditions under which GSRO belongs to $L_b(\Omega, X)$, the following definition is necessary.

Definition 2.2.3. The sequence $\{x_n;\ n \geq 1\}$ of elements of X is weakly 2-summable, if

$$\forall\, y \in X: \sum_{n=1}^{\infty}[x_n, y]^2 < +\infty.$$

Theorem 2.2.3. Let GSRO A have zero mean and the representation of theorem 2.2.1. Suppose that the set $MA^*A(X)$ is closed in X. Then $A \in L_b(\Omega, X)$ if, and only if, the measure W is discrete and the sequence of its weights, $\{W_k;\ k \geq 1\}$, satisfies the following conditions

$$\exists\, \Omega_0 \in \mathcal{F}:\ \mathrm{P}\,(\Omega_0) = 1,$$

$$\forall\, \omega\ \in \Omega_0 :\ \{\frac{1}{\sqrt{M|W_k|^2}} \cdot W_k(\omega);\ k \geq 1\} - \text{weakly 2-summable.}$$

Proof. *Necessity.* It follows from the conditions of the theorem that set $\sqrt{MA^*A}(X)$ is closed. In fact

$$\overline{MA^*A(X)} = MA^*A(X) \subset \sqrt{MA^*A(X)} \subset \overline{\sqrt{MA^*A(X)}} = MA^*A(X).$$

Denote $U_0 = \sqrt{MA^*A(X)}$. Since $U_0^* = U_0$, then U_0 is a bijection between $(\mathrm{Ker}\ U_0)^{\perp}$ and $U_0(X) = X_0$. Consequently, the operator U is a bijection between $(\mathrm{Ker}\ U_0)^{\perp}$ and $L_2([0;1],\lambda)$. Suppose that $A \in L_b(\Omega, X)$. Then $A^* \in L_s(\Omega, X)$. According to the previous theorem the measure W is discrete on $[0;1]$. Let $\{W_k;\ k \geq 1\}$ be the sequence of all weights of W which are indicated in an arbitrary order. The action of A on an element $x \in X$ can be written in the form

$$Ax = \sum_{k=1}^{\infty}[x, a_k] \cdot W_k = \sum_{k=1}^{\infty}[x, \sqrt{M|W_k|^2} \cdot a_k] \cdot \frac{W_k}{\sqrt{M|W_k|^2}},$$

where $\{a_k;\ k \geq 1\}$ is the sequence of elements of X and $[x, a_k]$ is the value of $(Ux) \in L_2([0;1],\lambda)$ at the point of the k-th jump of the measure W. Denote for every $k \geq 1$, $b_k := \sqrt{M|W_k|^2} \cdot a_k$. The map

$$X_0 \ni x \to \{[x, b_k];\ k \geq 1\} \in l_2 \tag{2.2.1}$$

is a bijection. Let the sequence $\{x_k;\ k \geq 1\}$ be such that

$$\forall\, j, k \geq 1 :\ [x_k, b_j] = \delta_{kj}.$$

Consider the family $\{\tilde{A}(\omega), \omega \in \Omega\}$ of bounded operators in X which corresponds to $A \in L_b(\Omega, X)$. Then

$$\forall\, k \geq 1 :\ (Ax_k)(\omega) = \frac{W_k(\omega)}{\sqrt{M|W_k|^2}} = \tilde{A}(\omega)x \quad (\mathrm{mod\ P}).$$

So there exists $\Omega_0 \in \mathcal{F}$, such that

$$\mathrm{P}(\Omega_0) = 1,$$

$$\forall \omega \in \Omega_0 :\ \forall\, k \geq 1 :\ \frac{W_k(\omega)}{\sqrt{M|W_k|^2}} = \tilde{A}(\omega)x_k.$$

It follows, from the Banach theorem, that map (2.2.1) has a continuous inverse. Hence

$$\forall\, x \in X_0 :\ x = \sum_{k=1}^{\infty}[x, b_k] \cdot x_k,$$

where the series converges in the norm of X. Consequently,

$$\forall\, \omega \in \Omega_0 : \ \forall\, x \in X_0 :$$

$$\tilde{A}(\omega)x = \sum_{k=1}^{\infty}[x, b_k] \cdot \frac{W_k(\omega)}{\sqrt{M|W_k|^2}}.$$

So, for every $\omega \in \Omega_0$, the map

$$l_2 \ni (\alpha_k;\ k \geq 1) \longrightarrow \sum_{k=1}^{\infty} \alpha_k \frac{W_k(\omega)}{\sqrt{M|W_k|^2}} \in X$$

is a bounded linear operator from l_2 to X. This condition is equivalent to the existence of the bounded adjoint operator.

$$X \ni y \to \left(\left[\frac{W_k(\omega)}{\sqrt{M|W_k|^2}}, y\right],\ k \geq 1\right) \in l_2. \tag{2.2.2}$$

It follows, from the closed graph theorem, that the bounded of (2.2.2) is equivalent to the condition of the weakly 2-summability of $\left\{\frac{W_k(\omega)}{\sqrt{M|W_k|^2}}, k \geq 1\right\}$. This establishes the claim.

Sufficiency. Let $\Omega_0 \in \mathcal{F}$ exist such that, for every $\omega \in \Omega_0$, the sequence $\left\{\frac{W_k(\omega)}{\sqrt{M|W_k|^2}}, k \geq 1\right\}$ is weakly 2-summable in X and $\mathrm{P}(\Omega_0) = 1$. Then the map (2.2.2) is a bounded linear map for all $\omega \in \Omega_0$. Hence, for all $\omega \in \Omega_0$, the map

$$l_2 \ni (\alpha_k;\ k \geq 1) \to \sum_{k=1}^{\infty} \alpha_k \frac{W_k(\omega)}{\sqrt{M|W_k|^2}} \in X$$

is a bounded linear map. Consequently, GSRO $A \in L_b(\Omega, X)$. Theorem is proved.

2.3 The integral representation of random maps

The main topic of this section is the structure of random maps in X. The goal is to develop the machinary for the consideration of equations with maps. The most famous examples of equations with stochastic maps are: the stochastic differential and the Ito-Volterra stochastic equations. In these equations the integral, with respect to a random measure, usually plays the role of a random map. So it is natural to try to represent the action of an arbitrary random map as an operation of stochastic integration. Since the random processes which arise in stochastic differential equations are measurable, with respect to a certain random measure, it is supposed, in the general case, that the random maps are measurable with respect to the generalized Gaussian random element from chapter 1. This condition allows the possibility to study random maps with the help of stochastic analysis. Let H and ξ be the same as in chapter 1, the real separable Hilbert space and the generalized Gaussian random element with zero mean and identity covariance operator. Suppose that $\mathcal{F} = \sigma(\xi)$. To consider the Ito-Wiener expansions of random maps the construction of X-valued multi-linear Hilbert-Schmidt forms from ξ are needed. Let $\{e_k;\ k \geq 1\}$ be an orthonormal basis in X and A_n an X-valued, n-linear symmetric Hilbert-Schmidt form on H. Define

$$A_n(\xi,\ldots,\xi) := \sum_{k=1}^{\infty}[A_n, e_k](\xi,\ldots,\xi)\cdot e_k, \tag{2.3.1}$$

where

$$\forall\, k \geq 1\ \forall\, \varphi_1,\ldots,\varphi_n \in H:$$
$$[A_n, e_k](\varphi_1,\ldots,\varphi_n) := [A_n(\varphi_1\ldots,\varphi_n), e_k].$$

It follows, from the properties of real-valued Hilbert-Schmidt forms from ξ, that: series (2.3.1) converges in X with probability 1; its sum does not depend on the choice of the basis $\{e_k;\ k \geq 1\}$ and

$$E|A_n(\xi,\ldots,\xi)|^2 < +\infty.$$

Moreover, X-valued forms have properties which are similar to properties of real-valued forms. Consider them without proofs. Denote by (H_n, X) the Hilbert space of all symmetric n-linear Hilbert-Schmidt X-valued forms on H. For $n = 0, (H_0, X) := X$. The inner product in (H_n, X) is defined in the following way

$$\forall\ A_n, B_n \in (H_n, X):$$

$$[A_n, B_n]_n := \sum_{t_1\ldots t_n=1}^{\infty} [A_n(\varphi_{t_1},\ldots,\varphi_{t_n}), B_n(\varphi_{t_1},\ldots,\varphi_{t_n})],$$

where $\{\varphi_t;\ t \geq 1\}$ is an arbitrary orthonormal basis in H. Then, for all $n, m \geq 0$, $A_n \in (H_n, X)$ and $B_m \in (H_m, X)$,
1) $EA_n(\xi,\ldots,\xi) = 0, n \geq 1$,
2) $E\,[A_n(\xi,\ldots,\xi), B_m(\xi,\ldots,\xi)] = 0$ if $n \neq m$,
and

$$E\,[A_n(\xi,\ldots,\xi), B_m(\xi,\ldots,\xi)] = n![A_n, B_n]_n.$$

The following statement about the Ito-Wiener expansion remains true.

Proposition. The random element x in X such that $E|x|^2 < +\infty$ has the unique expansion

$$x = \sum_{k=0}^{\infty} A_k(\xi,\ldots,\xi),$$

where $A_k \in (H_k, X), k \geq 0$ and the series converges in $L_2(\Omega, \mathrm{P}, X)$. Under this

$$E|x|^2 = \sum_{k=0}^{\infty} k!\|A_k\|_k^2.$$

Consider the random map F in X. For any $u \in X$, the random element $F(u)$ has the expansion

$$F(u) = \sum_{k=0}^{\infty} A_k(u)(\xi,\ldots,\xi).$$

It follows from the definition of random maps and properties of Hilbert-Schmidt forms from ξ, that for every $k \geq 0$, the map

$$X \ni u \longrightarrow A_k(u) \in (H_k, X)$$

is continuous. Consequently, the random map F can be defined by the sequence of non-random continuous maps from X to (H_k, X), $k \geq 0$. Consider certain examples.

Example 2.3.1. Let A be the GSRO in X which is jointly Gaussian with ξ. Then, for any $u \in X$, $A(u)$ is jointly Gaussian with ξ. So the Ito–Wiener expansion of $A(u)$ is

$$A(u) = \alpha_0(u) + \alpha_1(u)(\xi).$$

Here

$$\forall\, u \in X: \ \alpha_0(u) = EA(u).$$

The form α_1 is continuous and linear from every argument and, for any $u \in X$, $\alpha_1(u)(\cdot)$ is the Hilbert–Schmidt operator from H to X.

Example 2.3.2. Let $X = L_2([0;1])$, the random map such as in section 2.1.1. Consider $H = L_2([0;1])$ and define the generalized random element ξ as

$$\forall\, \varphi \in L_2([0;1]): \ (\varphi;\xi) = \int_0^1 \varphi(t)dw(t).$$

Then

$$\forall\, u \in X: \ F(u) = \beta_0(u) + \beta_1(u)(\xi),$$

where

$\forall\, \varphi \in H:$

$$\beta_0(u)(t) = u(t) - \int_0^t a(\tau, u(\tau))d\tau,$$

$$\beta_1(u)(\varphi)(t) = \int_0^t b(\tau, u(\tau))\varphi(\tau)d\tau.$$

Chapter 3

The composition of random maps

3.1 The reasons

The main aim of this chapter is to build the composition of random maps. Definitions will be used for solving equations with random maps. The composition of random maps requires an additional discussion because, in this situation, there are no "good" maps for every $\omega \in \Omega$. So the action of a random map on random elements is not well-defined. But, in the example from section 2.1.1, there exists a natural class of random functions which lies in the domain of the definition of the random map F.

This is the family of non-anticipating random functions with respect to the Wiener process w. As the action of F on these functions is defined with the help of the Ito stochastic integral, so it is natural to try to define the action of the random map on random elements, in the general case, with the help of the stochastic integral. The time evolvent for the elements in X is unnatural and can be done in different ways. So the integral to be used must be independent from the probabilistic relation of the random element and the random map. There are two ways of defining the anticipated stochastic integral. They are: the extended stochastic integral and the symmetric stochastic integral. In order to define the action of a random map on a random element, the extended stochastic integral is used.

3.2 Random multi-linear Hilbert–Schmidt forms of the generalized Gaussian random elements

The extended stochastic integral of chapter 1 can be regarded as an action of a random functional in H on a generalized element ξ. A multiple extended stochastic integral is defined in this section as an action of a random Hilbert-Schmidt form on ξ.

Let ζ be a random element in (H_k, X), such that $E\|\zeta\|^2 < +\infty$. Then, for every $\varphi_1, \ldots, \varphi_k \in H: \; E|\zeta(\varphi_1, \ldots, \varphi_k)|^2 < +\infty$. Consider the Ito-Wiener expansion

$$\zeta(\varphi_1, \ldots, \varphi_k) = \sum_{i=0}^{\infty} S_{ki}(\varphi_1, \ldots, \varphi_k, \xi, \ldots, \xi), \tag{3.2.1}$$

where $S_{ki}(\varphi_1, \ldots, \varphi_k, \ldots) \in (H_i, X), i \geq 0$. For every $i \geq 0$: S_{ki} is a Hilbert-Schmidt form from all $k+i$ arguments. For an orthonormal basis $\{\psi_t;\ t \geq 1\}$ in H

$$\sum_{t_1,\ldots,t_{k+i}=1}^{\infty} |S_{ki}(\psi_{t_1}, \ldots, \psi_{t_{k+i}})|^2 = \sum_{t_1,\ldots,t_k=1}^{\infty} \sum_{t_{k+1},\ldots,t_{k+i}=1}^{\infty} |S_{ki}(\psi_{t_1}, \ldots, \psi_{t_{k+i}})|^2 \leq$$

$$\sum_{t_1,\ldots,t_k=1}^{\infty} E|\zeta(\psi_{t_1}, \ldots, \psi_{t_k})|^2 = \frac{1}{i!} E\|\zeta\|_k^2 < +\infty.$$

Consequently, the symmetrization ΛS_{ki} of S_{ki} on all $k+i$ arguments is an element of (H_{k+i}, X).

Definition 3.2.1. If the series

$$\sum_{i=0}^{\infty} \Lambda S_{ki}(\xi, \ldots, \xi) \tag{3.2.1}$$

converges in the square mean in X, then its sum is called by the value of the random form ζ on ξ and denoted by $\zeta\langle\xi, \ldots, \xi\rangle$.

In the case $k = 1$, $X = R$ definition 3.2.1 coincides with the definition of an extended stochastic integral. Consider the example.

Example 3.2.1. Let $k = 2$, $X = R$, and $H = L_2(\mathcal{X}, \mathcal{A}, \mu)$, the generalized random element ξ is generated by Gaussian random measure γ, as in example 1.1.2. Now (H_2, R) is a symmetrical part of $L_2(\mathcal{X}^2, \mathcal{A} \times \mathcal{A}, \mu \times \mu)$. Consider disjoint $A, B, C \in \mathcal{A}$ and a random kernel (in this situation use 'kernel' instead of 'form')

$$\zeta(\theta_1, \theta_2) = \gamma(C) \cdot (\mathcal{X}_A(\theta_1) \cdot \mathcal{X}_B(\theta_2) + \mathcal{X}_B(\theta_1) \cdot \mathcal{X}_A(\theta_2)).$$

Then

$$\zeta(\theta_1, \theta_2) = (\mathcal{X}_A(\theta_1) \cdot \mathcal{X}_B(\theta_2) + \mathcal{X}_B(\theta_1) \cdot \mathcal{X}_A(\theta_2)) \cdot \int_C \gamma(d\theta_3) =$$

$$(\mathcal{X}_A(\theta_1) \cdot \mathcal{X}_B(\theta_2) + \mathcal{X}_B(\theta_1) \cdot \mathcal{X}_A(\theta_2)) \cdot \int_{\mathcal{X}} \mathcal{X}_C(\theta_3)\gamma(d\theta_3).$$

So the corresponding kernel

$$a_{12}(\theta_3, \theta_1, \theta_2) = \mathcal{X}_C(\theta_3)\,(\mathcal{X}_A(\theta_1) \cdot \mathcal{X}_B(\theta_2) + \mathcal{X}_B(\theta_1) \cdot \mathcal{X}_A(\theta_2)).$$

After symmetrization,

$$\Lambda a_{12}(\theta_3, \theta_1, \theta_2) = \frac{1}{3} \sum_{i=1}^{3} \mathcal{X}_C(\theta_i)\,(\mathcal{X}_A(\theta_{j_1}) \cdot \mathcal{X}_B(\theta_{j_2}) + \mathcal{X}_B(\theta_{j_1}) \cdot \mathcal{X}_A(\theta_{j_2})),$$

where $\{1, 2, 3\} = \{i, j_1, j_2\}$. Hence

$$\int\int_{\mathcal{X}^2} \zeta(\theta_1, \theta_2)\gamma(d\theta_1) \cdot \gamma(d\theta_2) = 2\gamma(A) \cdot \gamma(B) \cdot \gamma(C).$$

Sufficient condition for the convergence of (3.2.1) can be formulated in terms of stochastic derivatives of kernel ζ. Since (H_k, X) is a Hilbert space then, for the stochastic derivatives of random elements in (H_k, X), the definition from section 1.1 can be used. Let the random form $\zeta \in W^k(H_n, X)$. Then, for every $l = 1, \dots, k$, the stochastic derivative $\zeta^{(l)}$ is a Hilbert-Schmidt form from $n + k$ arguments and

$$\forall \varphi_1, \dots, \varphi_n, \psi_1, \dots, \psi_l :$$

$$\zeta^{(l)}(\varphi_1, \dots, \varphi_n; \psi_1, \dots, \psi_l) = D^l(\zeta(\varphi_1, \dots, \varphi_n))(\psi_1, \dots, \psi_l).$$

So $\zeta^{(l)}$ is symmetric on the first n arguments and the second l arguments separately. Denote for $l = 1, \dots, \min(k, n)$, $\varphi_1, \dots, \varphi_{l+n} \in H$

$$\zeta^{(l)}(\varphi_1, \dots, \varphi_{l+n})^* := \zeta^{(l)}(\varphi_{n+1}, \dots, \varphi_{n+l}, \varphi_{l+1}, \dots, \varphi_n; \varphi_1, \dots, \varphi_l).$$

For the r-linear X-valued Hilbert-Schmidt forms A, B on H, denote by $[A, B]_r$ their inner product which is defined similar to the inner product in (H_r, X).

Lemma 3.2.1. Let $k \geq 1, \zeta \in W^k(H_k, X)$. Then the value $\zeta < \xi, \dots, \xi >$ exists and

$$E|\zeta < \xi, \dots, \xi > |^2 = k!E\|\zeta\|_k^2 + \sum_{i=1}^{k} C_k^i \frac{k!}{i!} E[\zeta^{(i)}, \zeta^{(i)*}]_{k+i} \tag{3.2.3}$$

Proof. Consider the case when, for certain $j \geq 0$,

$$\forall \varphi_1, \dots, \varphi_k \in H : \ \zeta(\varphi_1, \dots, \varphi_k) = S_{kj}(\varphi_1, \dots, \varphi_k; \xi, \dots, \xi).$$

It follows, from the definition 3.2.1, that $\zeta < \xi, \dots, \xi >$ exists and

$$\zeta < \xi, \dots, \xi >= \Lambda S_{kj}(\xi, \dots, \xi).$$

Calculate $E|\zeta < \xi, \dots, \xi > |^2$. From the definition

$$\forall \varphi_1, \dots, \varphi_{k+j} \in H :$$

$$\Lambda S_{kj}(\varphi_1, \dots, \varphi_{k+j}) = \frac{1}{C_{k+j}^k} \sum_{1 \leq i_1 < \dots < i_k \leq k+j} S_{kj}(\varphi_{i_1}, \dots, \varphi_{i_k}; \varphi_{\nu_1}, \dots, \varphi_{\nu_j}),$$

where $\{\nu_1, \dots, \nu_j\} = \{1, \dots k + j\} \backslash \{i_1, \dots, i_k\}$. Denote by $\{\psi_t; \ t \geq 1\}$ an orthonormal basis in H. Then

$$E|\zeta < \xi, \dots, \xi > |^2 = E|\Lambda S_{kj}(\xi, \dots, \xi)|^2 =$$

$$= (k + j)! \|\Lambda S_{kj}\|_{k+j}^2 = (k + j)! \|\Lambda S_{kj}\|_{k+j}^2 =$$

$$= (k + j)! \sum_{t_1, \dots, t_{k+j}=1}^{\infty} |\Lambda S_{kj}(\psi_{t_1}, \dots, \psi_{t_{k+j}})|^2 =$$

$$= (k + j)! \frac{k!^2 \cdot j!^2}{(k + j)!^2} \cdot \sum_{l=1}^{k} C_{k+j}^k C_k^l C_j^l \times \sum_{t_1 \dots t_{k+j}=1}^{\infty} [S_{kj}(\psi_{t_1}, \dots, \psi_{t_k}; \ \psi_{t_{k+1}}, \dots, \psi_{t_{k+j}}),$$

$$S_{kj}(\psi_{t_{k+1}}, \dots, \psi_{t_{k+l}}, \psi_{t_{l+1}}, \dots, \psi_{t_k}; \ \psi_{t_1}, \dots, \psi_{t_l}, \psi_{t_{k+l+1}}, \dots, \psi_{t_{k+j}})] + k!E\|\zeta\|_k^2 =$$

$$= k!E\|\zeta\|_k^2 + \sum_{l=1}^{k} C_k^l \cdot \frac{k!}{l!} \cdot E[\zeta^{(l)}, \zeta^{(l)*}]_{k+l},$$

where C_j^l for $l > j$. Now for arbitrary $\zeta \in W^k(H_k, X)$ the claim can be established with the help of a limit procedure.

In the case $k = 1, X = R$ the statement of the lemma is known to be a sufficient condition for the existence of an extended stochastic integral. The conditions for the existence of a multiple extended stochastic integral are given in [25], but in this work there is no explicit expression for the second moment of such an integral.

Note that if, for the random element $\zeta \in (H_k, X)$, there exists $\zeta\langle\xi\ldots\xi\rangle$, then for every $i = 1,\ldots,k-1$, the random element $\xi_i = \zeta\langle\ldots,\xi \overset{i}{\ldots} \xi\rangle$ in (H_k, X) is well-defined and

$$\zeta\langle\xi\ldots\xi\rangle = \zeta_i\langle\underbrace{\xi\ldots\xi}_{k-i}\rangle.$$

Moreover, if $\zeta \in W^n((H_k, X))$, then $\zeta_i \in W^{n-i}((H_k, X))$.

The following theorem suggests that the multiple extended stochastic integral can be localized on smooth open sets.

Theorem 3.2.1. Let ζ_1, ζ_2 be random elements in (H_k, X) and values $\zeta_1\langle\xi\ldots\xi\rangle, \zeta_2\langle\xi\ldots\xi\rangle$ are well-defined. Suppose that G is such a smooth open set for which

$$\mathcal{X}_G \cdot \zeta_1 = \mathcal{X}_G \cdot \zeta_2 \qquad (\text{mod } P).$$

Then

$$\mathcal{X}_G \cdot \zeta_1\langle\xi\ldots\xi\rangle = \mathcal{X}_G \cdot \zeta_2\langle\xi\ldots\xi\rangle.$$

Proof. Use mathematical induction by k. In the case when $k = 1$, the statement of the theorem is theorem 1.2.1. Suppose that the claim is true for certain $k = k_0$. Consider random elements ζ_1, ζ_2 in (H_{k_0+1}, X) which satisfy the conditions of the theorem.

Since $(H_{k_0+1}, X) = (H_{k_0}, (H, X))$ then, due to the induction supposition random Hilbert-Schmidt operators $\zeta_1(\cdot, \xi, \overset{k_0}{\ldots} \xi)$ and $\zeta_2(\cdot, \xi, \overset{k_0}{\ldots} \xi)$ coincide on G a.s. with respect to P. Hence, due to theorem 1.2.1,

$$\mathcal{X}_G \cdot \zeta_1\langle\xi\ldots\xi\rangle = \mathcal{X}_G \cdot \zeta_2\langle\xi\ldots\xi\rangle \qquad (\text{mod } P).$$

Theorem is proved.

Theorem 3.2.1 allows such a definition to be formulated.

Definition 3.2.2. The random element ζ in (H_k, X) is termed k-times stochastically integrable, if there exist sequences $\{A_n; n \geq 1\}$ and $\{\zeta_n; n \geq 1\}$ of smooth open sets and random elements which satisfy definition 3.2.1 such that

$$1) \bigcup_{n=1}^{\infty} A_n = \Omega \qquad (\text{mod} P);$$

$$2) \forall n \geq 1 : \mathcal{X}_{A_n}\zeta_n = \mathcal{X}_{A_n}\zeta.$$

If the random element ζ is k-times stochastically integrable, then there exists a random variable α such that

$$\forall n \geq 1 : \mathcal{X}_{A_n} \cdot \zeta_n\langle\xi\ldots\xi\rangle = \mathcal{X}_{A_n} \cdot \alpha.$$

This random variable is called the multiple integral from ζ and is denoted by $\zeta\langle\xi\ldots\xi\rangle$

It follows from theorem 3.2.1 that such a statement is true.

Theorem 3.2.2. Every random element from $\tilde{W}^k((H_k, X))$ is k-times stochastically integrable.

Consider the following example.

Example 3.2.2. Let $H = L_2([0;1]), X = R$. The generalized random element ξ is defined by the equalities

$$\forall\varphi \in L_2([0;1]) : (\varphi, \xi) := \int_0^1 \varphi(t)dw(t),$$

where $\{w(t); t \in [0;1]\}$ is the Wiener process. In this case, $H_k = L_2^s([0;1]^k)$ is the symmetric subspace of $H = L_2([0;1]^k)$, $k \geq 1$. For $A_k \in H_k$,

$$A_k(\xi, \ldots, \xi) = k! \cdot \int_{\Delta_k(1)} a_k(t_1, \ldots, t_k)dw(t_1)\ldots dw(t_k),$$

$\Delta_k(1) = \{(t_1, \ldots, t_k) : 0 \leq t_1 \leq \ldots \leq t_k \leq 1\}$, where a_k is the corresponding kernel for A_k as in example 1.1.2′ . Consider the random kernel

$$a_2(t_1, t_2) = \exp(w(t_1)^n + w(t_2)^m), 0 \leq t_1 \leq t_2 \leq 1.$$

To prove that a_2 is twice stochastically integrable it is sufficient to verify that $a_2 \in \tilde{W}^2(L_2([0;1]^2))$. Consider the random element $\exp w(\cdot)^n$ in $L_2([0;1]), n \geq 1$. For every $n \geq 1, w(\cdot)^n$ has the stochastic derivative of all orders. Indeed, for $n = 1$, $w(\cdot)$ is a stochastically differentiable random element in $L_2([0;1])$ and its stochastic derivative has the following form

$$(Dw)\varphi = \int_0^{\cdot} \varphi(\tau)d\tau, \ \varphi \in L_2([0;1]).$$

Suppose that $w(\cdot)^k \in \bigcap_{j=1}^{\infty} W^j(L_2([0;1]))$ for $1 \leq k \leq n$. Then, from the Ito formula,

$$w(t)^{n+1} = \frac{1}{2}\int_0^t n(n-1)w(\tau)^{n-1}d\tau + \int_0^t (n+1)w(\tau)^n dw(\tau).$$

The first term in the right part is infinitely stochastically differentiable as an image under the bounded linear map of $w(\cdot)^{n-1}$. The second term is the value $\zeta\langle\xi\rangle$, for the form ζ

$$\zeta(\varphi) = \int_0^{\cdot} \varphi(\tau)(n+1)w(\tau)^n d\tau, \quad \varphi \in L_2([0;1]).$$

Hence, due to the remark to lemma 3.2.1, the second term has infinitely many stochastic derivatives. Now consider

$$a_2(t_1, t_2) = \left(\sum_{j_1=0}^{\infty} \frac{w(t_1)^{j_1 n}}{j_1!}\right)\left(\sum_{j_2=0}^{\infty} \frac{w(t_2)^{j_2 m}}{j_2!}\right).$$

The series from the right part converge with their derivatives of every order in the square mean on any set $\{\omega : \max_{[0;1]} |w(t)| < C\}$. Consequently, a_2 is infinitely stochastically differentiable and there exists the stochastic integral

$$\int_0^1 \int_0^{t_2} \exp(w(t_1)^n + w(t_2)^m)dw(t_1)dw(t_2).$$

3.3 The action of random map on random elements

Let F be a random map in X. Then there exists the family of non-random functions $\{f_k \in C(X,(H_k,X)); k \geq 0\}$ such that

$$\forall\, x \in X: \; F(x) = \sum_{k=0}^{\infty} f_k(x)(\xi,\ldots,\xi).$$

Definition 3.3.1. The random element η in X belongs to the domain of the definition of F if there exists the sequence $\{A_n;\; n \geq 1\}$ of smooth open sets such that, for every $k \geq 0$, the value $f_k(\eta) < \xi,\ldots,\xi >$ is defined with the help of a localization on $\{A_n;\; n \geq 1\}$ and the series

$$\sum_{k=0}^{\infty} f_k(\eta) < \xi,\ldots,\xi > \tag{3.3.1}$$

converges in the square-mean on every A_n, $n \geq 1$. The sum of (3.3.1) is the value $F(\eta)$.

Example 3.3.1. Let A and η be a GSRO and a random element in X respectively, which are commonly Gaussian with ξ. Then η belongs to the domain of the definition of A. Indeed

$$\forall\, x \in X: \; A(x) = \alpha_0(x) + \alpha_1(x)(\xi),$$

where $\alpha_0 \in L(X), \alpha_1 \in L(X,(H_1,X))$. The random element η now has the form

$$\eta = x_0 + Q(\xi),$$

where $Q \in (H_1, X)$. Consequently

$$A(\eta) = \alpha_0(x_0) + \alpha_0(Q(\xi)) +$$

$$+\sum_{i,j=1}^{\infty}(\alpha_1(Q(\psi_i))(\psi_j)\xi_i\psi_j - \delta_{ij}\alpha_1(Q(\psi_i))(\psi_j)).$$

Here $\{\psi_i;\; i \geq 1\}$ is an orthonormal basis in H, $\xi_i = (\psi_i,\xi)$, $i \geq 1$. Theorem 3.3.1 follows from theorem 3.2.1.

Theorem 3.3.1. Let F be a random map such that

$$\exists\, N \geq 1: \; \forall\, x \in X:$$

$$F(x) = \sum_{k=0}^{N} f_k(x)(\xi,\ldots,\xi).$$

Suppose that F is N-times continuously differentiable in the square mean, i.e. as a map from X to $L_2(\Omega, \mathrm{P}, X)$. Then every random element which has N stochastic derivatives belongs to the domain of F.

The proof follows from theorems 1.3.1 and 3.2.1.

Henceforth the random map F, which has a finite family $\{f_k\}$, will be called a polynomial from ξ. Consider certain properties of such random maps.

Theorem 3.3.1. Let F be a random map polynomial from ξ. Suppose that F satisfies Lipschitz condition in the square-mean. Consider the sequence

$$\{\eta_n;\ n \geq 1\} \subset L_2(\Omega, P, X)$$

of random elements such that for any k, n, the value $f_k(\eta_n)$ can be defined as in definition 3.2.1. Suppose, that for certain random elements $\eta, \zeta \in L_2(\Omega, \mathrm{P}, X)$,

$$E|\eta_n - \eta|^2 \to 0, E|F(\eta_n) - \zeta|^2 \to 0, n \to \infty,$$

$$\forall\ i = \overline{1, N}\ \exists\ L_2 - \lim_{n\to\infty} f_i(\eta_n) < \xi, \ldots, \xi > .$$

Then η belongs to the domain of the definition of F, and $F(\eta) = \zeta$.

Proof. It follows from the conditions of the theorem that for every $k = 0, \ldots, N$, f_k satisfies Lipschitz condition. Indeed

$$\forall\ k = 0, \ldots, N\ \forall x, y \in X :$$

$$\|f_k(x) - f_k(y)\|_k^2 \leq \frac{1}{k!} \cdot E|F(x) - F(y)|^2.$$

Since, for every $k = 0, \ldots, N$,

$$E\|f_k(\eta)\|_k^2 < +\infty.$$

Let

$$\forall\ k = 1, \ldots, N\ \ \forall\ \varphi_1, \ldots, \varphi_k \in H :$$

$$f_k(\eta_n)(\varphi_1, \ldots, \varphi_k) = \sum_{i=0}^{\infty} S_{ki}^n(\varphi_1, \ldots, \varphi_k, \xi, \ldots, \xi), n \geq 1,$$

$$f_k(\eta)(\varphi_1, \ldots, \varphi_k) = \sum_{i=0}^{\infty} S_{ki}(\varphi_1, \ldots, \varphi_k, \xi, \ldots, \xi).$$

Due to the conditions of the theorem

$$E|\eta_n - \eta|^2 \to 0, n \to \infty.$$

Hence

$$\forall\ k = 0, \ldots, N :$$

$$E\|f_k(\eta_n) - f_k(\eta)\|_k^2 \to 0, n \to \infty.$$

Consequently

$$\forall\ k = 1, \ldots, N\ \forall\ \varphi_1, \ldots, \varphi_k \in H :$$

$$\forall i \geq 0 :\ E|S_{ki}^n(\varphi_1, \ldots, \varphi_k, \xi, \ldots, \xi) - S_{ki}(\varphi_1, \ldots, \varphi_k, \xi, \ldots, \xi)|^2 =$$

$$= i!\|S_{ki}^n(\varphi_1, \ldots, \varphi_k, \cdot, \ldots, \cdot) - S_{ki}(\varphi_1, \ldots, \varphi_k, \cdot, \ldots, \cdot)\|_i^2 \to 0, n \to \infty.$$

So

$$\forall\ k = 1, \ldots, N\ \forall\ \varphi_1, \ldots, \varphi_k + i \in H :$$

$$|\Lambda S^n_{ki}(\varphi_1,\ldots,\varphi_{k+i}) - \Lambda S_{ki}(\varphi_1,\ldots,\varphi_{k+i}| \to 0, n \to \infty, \tag{3.3.2}$$

$$\forall\, i \geq 1\ \forall\, \varphi_1,\ldots,\varphi_i \in H:$$

$$|S^n_{0i}(\varphi_1,\ldots,\varphi_i) - S_{0i}(\varphi_1,\ldots,\varphi_i)| \to 0, n \to \infty,$$

$$|S^n_{00} - S_{00}| \to 0, n \to \infty,$$

where

$$f_0(\eta_n) = \sum_{i=0}^{\infty} S^n_{0i}(\xi,\ldots,\xi),$$

$$f_0(\eta) = \sum_{i=0}^{\infty} S_{0i}(\xi,\ldots,\xi).$$

Now

$$\zeta = \sum_{n=0}^{\infty} T_n(\xi,\ldots,\xi).$$

Since for every $k = 0,\ldots,N$ there exists

$$L_2 - \lim_{n\to\infty} f_k(\eta_n) < \xi,\ldots,\xi >,$$

then from (3.3.2) the series

$$f_k(\eta) < \xi,\ldots,\xi >= \sum_{i=0}^{\infty} \Lambda S_{ki}(\xi,\ldots,\xi)$$

converges in the square mean and

$$L_2 - \lim_{n\to\infty} f_k(\eta_n) < \xi,\ldots,\xi >= f_k(\eta) < \xi,\ldots,\xi > .$$

So η belongs to the domain of the definition of F and, due to the condition on ζ, $F(\eta) = \zeta$. Theorem is proved.

Theorem 3.3.2 states that a polynomial random map is 'closed'.

The action of the random map on random elements can be discontinuous, as in the following example.

Example 3.3.2. Let $X = H$. Consider such a random map F:

$$\forall x \in H: \ F(x) = (x,\xi)\cdot\varphi_0,$$

where $\|\varphi_0\| = 1$. For $n \geq 1$ define

$$\eta_n = \frac{1}{n}\sin[n(\varphi_1,\xi)]\cdot\varphi_1,$$

where $\varphi_1 \in H, \|\varphi_1\| = 1$. Then (cf. example 1.1.2)

$$E\|\eta_n\|^2 \to 0, n \to \infty,$$

$$\forall\, n \geq 1:$$

$$F(\eta_n) = \left\{ \frac{1}{n} \sin[n(\varphi_1, \xi)] \cdot (\varphi_1, \xi) - \cos[n(\varphi_1, \xi)] \right\} \cdot \varphi_0,$$

$$E\|F(\eta_n)\|^2 = \frac{1}{n^2} \frac{1}{\sqrt{2\pi}} \int_R \sin^2(nx) \cdot x^2 \cdot e^{-\frac{x^2}{2}} dx -$$

$$-\frac{2}{n} \frac{1}{\sqrt{2\pi}} \cdot \int_R \sin(nx) \cdot \cos(nx) \cdot x e^{-\frac{x^2}{2}} dx +$$

$$+\frac{1}{\sqrt{2\pi}} \cdot \int_R \cos^2(nx) \cdot e^{-\frac{x^2}{2}} dx \to \frac{1}{2},\ n \to \infty.$$

Definition 3.3.1 allows the superposition of random maps to be defined. Consider the case of linear maps.

Lemma 3.3.1. Let A, B be polynomial random linear operators in X, i.e.

$$\forall\, x \in X: \ Ax = \sum_{k=0}^{L} a_k(x)(\xi, \dots, \xi),$$

$$Bx = \sum_{j=0}^{N} b_j(x)(\xi, \dots, \xi),$$

where $a_k, b_j \in L(X, (H_k, X))\,(L(X, (H_j, X)))$, $k = 0, \dots, L$, $j = 0, \dots, N$. Then for every $x \in X$ a random element Ax belongs to the domain of the definition of B and the superposition $B \circ A$ is a random linear map.

Proof. The part of the theorem which states, that Ax belongs to the domain of the definition of B is a consequence of theorem 3.3.1. To prove that $B \circ A$ is continuous in the square-mean, use the equality from theorem 3.2.1. Note that for every $i = 1, \dots, N$,

$$\forall\, l = 1, \dots, i:$$

$$E\|D^l b_i(\sum_{k=0}^{L} a_k(x)(\xi, \dots, \xi))\|_{i+l}^2 \leq$$

$$\leq L!\|b_i\|^2 \cdot E\|\sum_{k=0}^{L} f_k(x)(\xi, \dots, \xi)\|^2 =$$

$$= L!\|b_i\|^2 \cdot E\|Ax\|^2,$$

where $\|b_i\|$ is the operator norm of a linear operator $b_i \in L(X, (H_i, X))$. The claim is proved.

Consequence. Let $A_1, \dots, A_n$ be GSROs in H which are jointly Gaussian with ξ. Then the superposition $A_n \circ \dots \circ A_1$ exists and does not depend on the arrangement of brackets.

3.4 Bounded random operators

Consider $A, B \in L_b(\Omega, X)$. Denote by $\{A(\omega), \omega \in \Omega\}$ and $\{B(\omega), \omega \in \Omega\}$ corresponding families of continuous operators. Then the superposition of A and B can be defined thus for every ω:

$$B \times A(\omega) := B(\omega)A(\omega), \ \omega \in \Omega.$$

The result of such a superposition is a bounded random operator in X [26]. In certain cases, the superposition $B \circ A$ can be defined as in section 3.3. Generally speaking, $B \times A \neq B \circ A$.

Example 3.4.1. Let $A = \eta \cdot A_0$, $B = \eta \cdot B_0$, where $A_0, B_0 \in L(X)$ and η is jointly Gaussian with ξ. Suppose that η has zero mean and variance 1. Then the superposition for every ω is

$$B \times A = \eta^2 \cdot A_0 B_0,$$

and the superposition in the sense of section 3.3

$$B \circ A = (\eta^2 - 1) B_0 A_0.$$

Consider the case when $B = \zeta \cdot B_0$, where ζ is also jointly Gaussian with ξ but independent from η. Then

$$B \times A = B \circ A = \eta \cdot \zeta \cdot B_0 A_0.$$

The main aim of this section is to find the connection between $B \times A$ and $B \circ A$. To do this, it is enough to find the connection between the random elements $A(\omega)\eta(\omega)$, $\omega \in \Omega$ and $A\eta$ in X, where $A\eta$ is the result of the action of A on η in the sense of section 3.3. The following definition will be necessary.

Let A be a linear random map in X. Then

$$\forall\, x \in X: \ Ax = \sum_{k=0}^{\infty} \alpha_k(x)(\xi, \ldots, \xi),$$

where $\alpha_k \in L(X, (H_k, X)), k \geq 0$.

Definition 3.4.1. A random operator A is called regular if
1) $A \in L_b(\Omega, X)$
2) for any $k \geq 1$ the random operator

$$X \ni x \to \alpha_k(x)(\xi, \ldots, \xi)$$

is bounded.

Note that an arbitrary bounded random operator may not be regular.

Example 3.4.2. Let $\{e_i;\ i \geq 1\}$, $\{\varphi_i;\ i \geq 1\}$ be orthonormal bases in X and H respectively. Denote by $\xi_i = (\varphi_i; \xi)$, $i \geq 1$. Consider the random operator

$$\forall x \in X: \ Ax := \sum_{i=1}^{\infty} \operatorname{sign} \xi_i \cdot [x; e_i] \cdot e_i.$$

Then $A \in L_b(\Omega, X)$. Moreover A is an isometry with probability 1. Consider the following Ito–Wiener expansion of A:

$$Ax = \sum_{k=0}^{\infty} \alpha_k(x)(\xi, \ldots, \xi).$$

Find $\alpha_1(\cdot)(\xi)$. It follows from the properties of Hilbert–Schmidt forms from ξ that, for arbitrary $x, y \in X$, the random variable $[\alpha_1(x)(\xi), y]$ is a projection in $L_2(\Omega, \mathcal{F}, \mathrm{P})$ of a random variable $[Ax, y]$ on the subspace $(\varphi, \xi); \varphi \in H$. Hence

$$\forall\, x \in X : \alpha_1(x)(\xi) = \sum_{i=1}^{\infty} (M|\xi_i|) \cdot \xi_i \cdot [x, e_i] e_i.$$

Since

$$\forall\, i \geq 1 : M|\xi_i| = \sqrt{\frac{2}{\pi}}\,,$$

then $\alpha_1(\cdot)(\xi) \bar{\in} L_b(\Omega, X)$.

Consequently A is not regular.

Consider $\alpha_k \in (H_k, X)$ and $\beta_l \in (H_l, X)$. For any $i = 1, \ldots, min(k, l)$ denote by $(\alpha_k, \beta_l)_{\sigma i}$ the multilinear Hilbert–Schmidt form $\Lambda\kappa$ from ξ , where $\forall \psi_1, \ldots, \psi_{k+l-2i} \in H$:

$$\kappa(\psi_1, \ldots, \psi_{k+l-2i}) := \frac{k!}{(k-i)!} \cdot \frac{l!}{(l-i)!} \times$$

$$\times \sum_{s_1 \ldots s_i = 1}^{\infty} [\alpha_k(\psi_1, \ldots, \psi_{k-i}, \varphi_{s_1}, \ldots, \varphi_{s_i}), \beta_l(\psi_{k-i+1}, \ldots, \psi_{k+l-2i}, \varphi_{s_1}, \ldots, \varphi_{s_i})].$$

Here $\{\varphi_s; s \geq 1\}$ is an orthonormal basis in H. The random variable $(\alpha_k, \beta_l)_{\sigma i}$ can be expressed through the random variables

$$[D^j \alpha_k(\xi, \ldots, \xi), D^j \beta_l(\xi, \ldots, \xi)]_j, j \leq 1.$$

Lemma 3.4.1. For arbitrary $k, l \geq 1, \alpha_k \in (H_k, X), \beta_l \in (H_k, X)$

$\forall i = 0, \ldots, \min(k, l)$:

$$(\alpha_k, \beta_l)_{\sigma i} = \sum_{j=0}^{\min(k,l)-i} \frac{(-1)^j}{j!} (D^{i+j} \alpha_k, D^{i+j} \beta_l)_{i+j},$$

where $(\alpha_k, \beta_l)_{\sigma o}$ is a Hilbert–Schmidt form from ξ, which is a symmetrization of $[\alpha_k, \beta_l] : H^{k+l} \to R$.

Proof. It follows, from the work [25], that for arbitrary $\gamma_k \in H_k, \delta_l \in H_l$:

$$\gamma_k(\xi, \ldots, \xi)\delta_l(\xi, \ldots, \xi) = \sum_{i=0}^{\min(k,l)} \frac{1}{i!} (\gamma, \delta_l)_{\sigma i}.$$

Using this formula it is possible to obtain $\forall k, l \geq 1 \;\; \forall \alpha_k \in (H_k, X), \;\; \beta_l \in (H_l, X)$:

$$[\alpha_k(\xi, \ldots, \xi), \beta_l(\xi, \ldots, \xi)] = \sum_{i=0}^{\min(k,l)} \frac{1}{i!} (\alpha_k, \beta_l)_{\sigma i}.$$

An analogous equality for $[D^i\alpha_k(\xi,\dots,\xi), D^i\beta_l(\xi,\dots,\xi)]_i$ $i = 1,\dots,\min(k,l)$ is true:

$$[D^i\alpha_k(\xi,\dots,\xi), D^i\beta_l(\xi,\dots,\xi)]_i = \sum_{i=0}^{\min(k,l)-i} \frac{1}{i!}(\alpha_k,\beta_l)_{\sigma i+j}.$$

Consider these equations as a system on $(\alpha_k,\beta_l)_{\sigma i}$, $i = 0,\dots,\min(k,l)$. Now to prove the claim it is enough to solve the system.

Such a statement follows from lemma 3.4.1.

Theorem 3.4.1. Let A be a regular random operator in X such that

$$\forall x \in X \ : Ax = \sum_{k=0}^{N} \alpha_k(x)(\xi,\dots,\xi),$$

where $\alpha_k \in L(X,(H_k,X))$, $k = 0,\dots,N$. Let η be a random element in X, which has N stochastic derivatives. Then the result of the action of A on η, in the sense of section 3.3, is such that $A\eta$ can be written in the form

$$(A\eta)(\omega) = A(\omega)\eta(\omega) + \sum_{s=1}^{\infty}\Big\{\sum_{i=1}^{N}\frac{(-1)^i}{i!}\sum_{t_1\dots t_i=1}^{\infty} D^iA(e_s)(\varphi_{t_1},\dots,\varphi_{t_i})(\omega)\times$$

$$\times[D^i\eta(\varphi_{t_1},\dots,\varphi_{t_i})(\omega), e_s]\Big\} \qquad (\mathrm{mod}\ P),$$

where $\{e_s; s \geq 1\}, \{\varphi_t; t \geq 1\}$ are orthonormal bases in X and H. The multiple series converge with probability 1 and the series indicated by s converges in the square-mean.

Proof. Consider first the bounded random operator $A_k = \alpha_k(\cdot)(\xi,\dots,\xi)$. Let $\eta = \sum_{l=0}^{\infty} S_l(\xi,\dots,\xi)$.

Denote by γ_l, γ_{ls} for $l \geq 0, s \geq 1$ the following $k+l$ linear forms on X:

$$\forall\psi_1,\dots,\psi_{k+l} \in H :$$

$$\gamma_l(\psi_1,\dots,\psi_{k+l}) := \alpha_k(S_l(\psi_1,\dots,\psi_l))(\psi_{l+1},\dots,\psi_{l+k}),$$

$$\gamma_{ls}(\psi_1,\dots,\psi_{k+l}) := \alpha_k(e_s)(\psi_1,\dots,\psi_{l+k})[S_l(\psi_1,\dots,\psi_l), e_s].$$

Note that

$$|\sum_{s=1}^{n} \Lambda\gamma_{ls} - \Lambda\gamma_l|_{k+l} \to 0, n \to \infty.$$

Indeed,

$$|\sum_{s=1}^{n} \Lambda\gamma_{ls} - \Lambda\gamma_l|_{k+l}^2 \leq |\sum_{s=1}^{n} \gamma_{ls} - \gamma_l|_{k+l}^2 =$$

$$= \frac{1}{l!}\ E|\alpha_k(\sum_{s=1}^{n}[S_l(\xi,\dots,\xi), e_s]\cdot e_s - S_l(\xi,\dots,\xi))|_k^2 \leq$$

$$\frac{1}{l!}\ \|\alpha_k\|^2\cdot E|\sum_{s=1}^{n}[\eta, e_s]\cdot e_s - \eta|^2 \to 0,\ n \to \infty,$$

where $\|\alpha_k\|$ is the operator's norm of $\alpha_k : X \to (H_k, X)$. Consequently

$$\alpha_k(\eta)\langle\xi, \ \dots \ ,\xi\rangle = \sum_{l=0}^{\infty} \Lambda\gamma_l(\xi, \overset{k+l}{\dots}, \xi) = \sum_{s=1}^{\infty}\sum_{l=1}^{\infty} \Lambda\gamma_{ls}(\xi, \ \dots \ ,\xi).$$

It follows from lemma 3.4.1, that

$$\forall l \geq 0, s \geq 1 : \ \ \Lambda\gamma_{ls}(\xi, \dots, \xi) =$$

$$= \sum_{i=0}^{\min(k,l)} \frac{(-1)^i}{i!} \sum_{t_1 \dots t_i = 1}^{\infty} D^i\alpha_k(e_s)(\xi, \dots, \xi)(\varphi_{t_1}, \dots, \varphi_{t_i}) \cdot$$

$$\cdot [D^i S_l(\xi, \dots, \xi)(\varphi_{t_1}, \dots, \varphi_{t_i}), e_s] =$$

$$= \sum_{i=0}^{N} \sum_{t_1 \dots t_i = 1}^{\infty} \frac{(-1)^i}{i!} D^i\alpha_k(e_s)(\xi, \dots, \xi)(\varphi_{t_1}, \dots, \varphi_{t_i}) \cdot$$

$$\cdot [D^i S_l(\xi, \dots, \xi)(\varphi_{t_1}, \dots, \varphi_{t_i}), e_s].$$

Hence

$$\alpha_k(\eta)\langle\xi, \ \dots \ ,\xi\rangle = \sum_{s=1}^{\infty}\sum_{l=0}^{\infty}\sum_{i=0}^{N}\sum_{t_1 \dots t_i = 1}^{\infty} \frac{(-1)^i}{i!} \cdot$$

$$D^i\alpha_k(e_s)(\xi, \dots, \xi)(\varphi_{t_1}, \dots, \varphi_{t_i}) \cdot [D^i S_l(\xi, \dots, \xi)(\varphi_{t_1}, \dots, \varphi_{t_i}), e_s] =$$

$$= \sum_{s=1}^{\infty}\sum_{i=0}^{N}\sum_{t_1 \dots t_i = 1}^{\infty} \frac{(-1)^i}{i!} D^i\alpha_k(e_s)(\xi, \dots, \xi)(\varphi_{t_1}, \dots, \varphi_{t_i}) \cdot$$

$$\cdot [D^i \eta(\varphi_{t_1}, \dots, \varphi_{t_i}), e_s] = \sum_{s=1}^{\infty} \alpha_k(e_s)(\xi, \dots, \xi) \cdot [\eta, e_s] +$$

$$+ \sum_{s=1}^{\infty}\sum_{i=1}^{N}\sum_{t_1 \dots t_i = 1}^{\infty} \frac{(-1)^i}{i!} D^i\alpha_k(e_s)(\xi, \dots, \xi)(\varphi_{t_1}, \dots, \varphi_{t_i}) \cdot$$

$$\cdot [D^i \eta(\varphi_{t_1}, \dots, \varphi_{t_i}), e_s] = \alpha_k(\eta)(\xi, \dots, \xi) +$$

$$+ \sum_{s=1}^{\infty}\sum_{i=1}^{N}\sum_{t_1 \dots t_i = 1}^{\infty} \frac{(-1)^i}{i!} D^i\alpha_k(e_s)(\xi, \dots, \xi)(\varphi_{t_1}, \dots, \varphi_{t_i}) \cdot$$

$$\cdot [D^i \eta(\varphi_{t_1}, \dots, \varphi_{t_i}), e_s],$$

where $\alpha_k(\eta)(\xi, \dots, \xi)$ is the result of the action of the bounded random operator $\alpha_k(\cdot)(\xi, \dots, \xi)$ on a random element η at every ω. Now, to finish the proof, it is enough to add the same equalities for $k = 0, \dots, N$.

Example 3.4.3. *Stratonovich (symmetric) stochastic integral.* Let $X = H = L_2(\mathcal{X}, \mathcal{A}, \mu)$, and the generalized Gaussian random element ξ is defined with the help of the Gaussian random measure θ as in example 1.1.2. Consider a random functional in $L_2(\mathcal{X}, \mathcal{A}, \mu)$ such that

$$\forall \varphi \in L_2(\mathcal{X}, \mathcal{A}, \mu) : \mathrm{F}(\varphi) := \int_{\mathcal{X}} \varphi(\mathrm{t})\theta(\mathrm{dt}).$$

It has been mentioned above and can be easily checked that, in the case of dim $L_2(\mathcal{X}, \mathcal{A}, \mu) = \infty$, F is not a bounded random functional. But, due to the linearity of F, the action of F on random elements may be defined similarly to a bounded case. Consider an orthonormal basis $\{\varphi_k; k \geq 1\}$ in $L_2(\mathcal{X}, \mathcal{A}, \mu)$. Define for the random element η, with the finite second moment,

$$F(\eta) = \int_{\mathcal{X}} \eta(t) \circ \theta(dt) := \sum_{k=1}^{\infty} \int_{\mathcal{X}} \varphi_k(t)\theta(dt) \cdot \int_{\mathcal{X}} \varphi_k(t)\eta(t)\mu(dt),$$

if the series converges in the square-mean. This definition coincides with the definition of the symmetric (Stratonovich in particular case) integral from section 1.4. The equality (3.4.1) is similar to the action "at every ω " and coincides with it when F is a bounded random functional with an essentially bounded norm. Consider a random element such as η, which has a nuclear stochastic derivative, i.e. for which $D\eta$ is not only a random element in (H, H) but is also a nuclear operator in H with probability 1, and $D\eta$ has finite second moment in the nuclear norm. It follows from theorem 3.4.1, that

$$\forall\, N \geq 1 : \sum_{k=1}^{N} \int_{\mathcal{X}} \varphi_k(t)\theta(dt) \cdot \int_{\mathcal{X}} \varphi_k(t) \cdot \eta(t)\mu(dt) =$$

$$= \sum_{k=1}^{N} \int \int_{\mathcal{X}} D\eta(t,s)\varphi_k(t)\varphi_k(s)\mu(dt)\mu(ds) +$$

$$+ \int_{\mathcal{X}} \left\{ \sum_{k=1}^{N} \int_{\mathcal{X}} \varphi_k(t)\eta(t)\mu(dt) \cdot \varphi_k(s) \right\} \theta(ds),$$

where the second term in the right-hand side is an extended stochastic integral. Due to the condition on η

$$L_2 - \lim_{N\to\infty} \sum_{k=1}^{N} \int \int_{\mathcal{X}} D\eta(t,s)\varphi_k(t)\varphi_k(s)\mu(dt)\mu(ds) = \text{tr } D\eta,$$

$$L_2 - \lim_{N\to\infty} \int_{\mathcal{X}} \left\{ \sum_{k=1}^{N} \int_{\mathcal{X}} \varphi_k(t)\eta(t)\mu(dt) \cdot \varphi_k(s) \right\} \theta(ds) = \int_{\mathcal{X}} \eta(s)\theta(ds).$$

Consequently, the symmetric stochastic integral from η exists, does not depend on the choice of the basis $\{\varphi_k;\ k \geq 1\}$ and

$$\int_{\mathcal{X}} \eta(t) \circ \theta(dt) = \int_{\mathcal{X}} \eta(t)\theta(dt) + \text{tr } (D\eta). \qquad (3.4.2)$$

Note that this relation between a symmetric and an extended stochastic integrals is well-known (cf. for example [27]). For example, due to (3.4.2)

$$\int_0^1 w(t) \circ dw(t) = \int_0^1 w(t)dw(t) + \frac{1}{2} = \frac{1}{2}w(1)^2.$$

Indeed w, as a random element in $L_2([0;1])$, is stochastically differentiable and

$$Dw(t,s) = \chi_{[0;t]}(s).$$

So

$$\text{tr } Dw = 2\sum_{n=1}^{\infty}\int_0^1 \sin(n\pi t)\cdot\int_0^t \sin(n\pi s)dsdt =$$

$$= \frac{4}{\pi^2}\cdot\sum_{k=0}^{\infty}\frac{1}{(2k+1)^2} = \frac{1}{2}.$$

Theorem 3.4.1 has the following consequence.

Consequence. Let A, B be the regular operators which satisfy the conditions of theorem 3.4.1. Suppose that A and B are independent in the following sense: there are orthogonal subspaces M and L in H, such that A is measurable with respect to $\sigma(\{(\varphi,\xi);\ \varphi\in M\})$ and B is measurable with respect to $\sigma(\{(\varphi,\xi);\ \varphi\in L\})$. Then the superposition AB in the sense of section 3.3 is equal to $A(\omega)B(\omega), \omega\in\Omega$.

3.5 Generalized Gaussian functionals of the first kind

It has been mentioned in the previous section that the different actions of random linear operators on random elements are closely connected with the different stochastic integrals. In the following two sections the generalized Fourier transformation of Wiener functionals will be constructed. This transformation transforms an extended stochastic integral into a symmetric stochastic integral. For this consideration a notion of generalized Gaussian functionals will be needed. Two constructions of such functionals will be considered.

Existence of the Ito-Wiener expansion allows the space $L_2(\Omega,\mathcal{F},\mathrm{P})$ (henceforth simply L_2) to be identified as the space $\bigoplus\sum_{k=0}^{\infty}\sqrt{k!}H_k$. Let $\{r_k;\ k\geq 0\}$ be a sequence of positive numbers such that

$$\sup_{k\geq 0} r_k < +\infty.$$

Definition 3.5.1. The space $\oplus\sum_{k=0}^{\infty}\sqrt{k!}r_k H_k$ is called by the space L_2^- of generalized Gaussian functionals of the first kind. The space $\oplus\sum_{k=0}^{\infty}\sqrt{k!}\cdot r_k^{-1}\cdot H_k$ is called by the space L_2^+ of test Gaussian functionals of the first kind.

Note that L_2^- and L_2^+ depend on the choice of the sequence $\{r_k;\ k\geq 0\}$. It follows, from the condition on $\{r_k;\ k\geq 0\}$, that the space of test functionals L_2^+ can be embedded into L_2 if every sequence $(A_k;\ k\geq 0)\in L_2^+$ can be identified with the random variable

$$\sum_{k=0}^{\infty} A_k(\xi,\ldots,\xi).$$

In an analogous way $L_2\subset L_2^-$. Moreover it can be said that $(L_2^+)^* = L_2^-, (L_2^-)^* = L_2^+$ if for $\alpha^+ = (A_k;\ k\geq 0)\in L_2^+$ and $\beta^- = (B_k;\ k\geq 0)\in L_2^-$ the following is defined,

$$<\alpha^+,\beta^->:=\sum_{k=0}^{\infty} k!(A_k,B_k)_k.$$

Since $L_2^+ \subset L_2$ as the space of usual test functions $\mathcal{D}(R^n) \subset L_2(R^n)$, then it can be expected that certain elements of L_2^-, such as the regular generalized functions from D^*, are usual random variables. The following example is an argument in favour of this supposition.

Example 3.5.1. Consider the case $H = R$. Here the generalized random element ξ is simply a random variable. The following equation

$$x - I(x) = 1 \tag{3.5.1}$$

is now an equation for random variables. Suppose that there exists a solution $x \in D$. Then

$$x = \sum_{k=0}^{\infty} a_k \overline{H}_k(\xi),$$

$$I(x) = \sum_{k=0}^{\infty} a_k \overline{H}_{k+1}(\xi).$$

It follows from (3.5.1) that

$$a_k = 1, k \geq 0.$$

But the series

$$\sum_{k=0}^{\infty} \overline{H}_k(\xi)$$

does not converge in L_2 .So we can say that (3.5.1) has a generalized solution which is an element of L_2^-. Hence L_2^- is built with the help of the sequence $\{r_k = \frac{1}{(k!)^\delta};\ k \geq 0\}$ for certain fixed $\delta > 1/2$. On the other hand if equation (3.5.1) is considered after localization of I the smooth solution $x = f(\xi)$ can be found with the help of the usual differential machinery. Indeed, as in chapter 1, for $f \in C^1(R)$

$$I(f(\xi)) = f(\xi) \cdot \xi - f'(\xi).$$

So (3.5.1) is a differential equation

$$f(\xi) - f(\xi) \cdot \xi + f'(\xi) = 1, \tag{3.5.2}$$

which has a non-empty set of solutions. Hence the generalized functional $\{\overline{H}_k(\xi);\ k \geq 0\}$ can be identified with the certain random variable $f_0(\xi)$, where f_0 is a solution of (3.5.2).

To define the set of a regular generalized functional use the Fourier-Wiener transformation. Henceforth this section we suppose that $\dim H < +\infty$, i.e. $H = R^N$ for fixed $N \geq 1$. The generalized element ξ is now the random vector $(\xi_1, \ldots, \xi_N)$ with independent identically-distributed coordinates. For $\alpha \in L_2$ with the expansion

$$\alpha = \sum_{k=0}^{\infty} \sum_{t_1 \ldots t_k = 1}^{N} c_{t_1 \ldots t_k} \cdot \xi_{t_1} * \cdots * \xi_{t_k} = f(\xi_1, \ldots, \xi_N)$$

define [28, 29]

$$\mathcal{T}(\alpha)(\lambda_1, \ldots, \lambda_N) = \sum_{k=0}^{\infty} i^k \sum_{t_1 \ldots t_k = 1}^{N} c_{t_1 \ldots t_k} \lambda_{t_1} \cdot \ldots \cdot \lambda_{t_k} =$$

$$= \exp(\frac{1}{2}\sum_{t=1}^{N}\lambda_t^2)\cdot E\alpha\cdot\exp(i\cdot\sum_{t=1}^{N}\lambda_t\xi_t) =$$

$$= \exp(\frac{1}{2}\sum_{t=1}^{N}\lambda_t^2)\cdot\frac{1}{(\sqrt{2\pi})^N}\cdot$$

$$\cdot\int_{R^N} f(u_1,\ldots,u_N)\cdot\exp(-\frac{1}{2}\sum_{t=1}^{N}u_t^2 + i\sum_{t=1}^{N}\lambda_t u_t)du,$$

$$u = (u_1,\ldots,u_N), \lambda = (\lambda_1,\ldots,\lambda_N)\in R^N. \qquad (3.5.3)$$

Hence,

$$\mathcal{T}(\alpha)(\lambda) = \exp(\frac{1}{2}\|\lambda\|^2)\cdot(f(u)\cdot\exp(-\frac{1}{2}\|u\|^2))^{v}(\lambda),$$

where v is the symbol of the Fourier transform and $\|\cdot\|$ is the Euclidean norm in R^N. Consequently, every random variable $\alpha\in L_2$ is uniquely determined by its transformation $\mathcal{T}(\alpha)$ and

$$\alpha = \exp(\frac{1}{2}\|\xi\|^2)\cdot(\tau(\alpha)(\lambda)\cdot\exp(-\frac{1}{2}\|\lambda\|^2))^{\wedge}(\xi),$$

where $\wedge$ is the symbol of the inverse Fourier transformation. The Fourier-Wiener transformation is not an isometry in L_2 but, for arbitrary $\alpha,\beta\in L_2$, the mathematical expectation $E\alpha\beta$ can be obtained in terms of its τ-transformation. Let

$$\beta = g(\xi) = g(\xi_1,\ldots,\xi_N).$$

Then

$$E\alpha\beta = \frac{1}{(\sqrt{2\pi})^N}\cdot\int_{R^N} f(u)g(u)\exp(-\frac{1}{2}\|u\|^2)du =$$

$$\sum_{k=0}^{\infty}k!\sum_{t_1\ldots t_k=1}^{N} c_{t_1\ldots t_k}d_{t_1\ldots t_k},$$

where $\{(d_{t_1\ldots t_k})_1^N;\ k\geq 0\}$ are elements of the Ito-Wiener expansion of β. Denote $\mathcal{T}(\alpha)=\varphi$, $\mathcal{T}(\beta)=\psi$. It follows, from (3.5.3), that φ and ψ can be extended to the analytical functions on C^N. Consequently

$$E\alpha\beta = \sum_{k=0}^{\infty}k!\frac{1}{(2\pi i)^N}\cdot\sum_{t_1\ldots t_k=1}^{N}\frac{(-1)^k}{s_1!\cdot\ldots\cdot s_N!}\cdot$$

$$\int_{\Gamma_R}\frac{\psi(\lambda)}{\lambda_{t_1}\ldots\lambda_{t_k}}\cdot\frac{\partial^k\varphi(0)}{\partial\lambda_{t_1}\ldots\partial\lambda_{t_k}}\cdot\frac{1}{\lambda_1\ldots\lambda_N}d\lambda_1\ldots d\lambda_N, \qquad (3.5.4)$$

$$\Gamma_R = \{(\lambda_1,\ldots,\lambda_N):\ |\lambda_1| = |\lambda_2| = \ldots = |\lambda_N| = R\}.$$

Here $\lambda_1^{s_1},\ldots,\lambda_N^{s_N} = \lambda_{t_1},\ldots,\lambda_{t_k}$. If $\alpha\in L_2^+$ then

$$\forall\, k\geq 0\ \ \forall\, t_1\ldots t_k = 1,\ldots,N$$

$$c_{t_1\ldots t_k}^2 \leq \frac{L}{k!^{1+2\delta}},$$

where $L = \|\alpha\|_+^2$. Hence, for any positive $\delta_1, \ldots, \delta_N$,

$$\sum_{k=0}^{\infty} \sum_{t_1 \ldots t_k=1}^{N} \left| \frac{\partial^k \varphi(0)}{\partial \lambda_{t_1} \ldots \partial \lambda_{t_k}} \right| \delta_{t_1}, \ldots, \delta_{t_k} =$$

$$= \sum_{k=0}^{\infty} \sum_{t_1 \ldots t_k=1}^{N} |c_{t_1 \ldots t_k}| \cdot s_1! \cdot \ldots \cdot s_N! \delta_{t_1} \cdot \ldots \cdot \delta_{t_k} =$$

$$= \sum_{k=0}^{\infty} k! \sum_{t_1 \ldots t_k=1}^{N} |c_{t_1 \ldots t_k}| \cdot \frac{s_1! \cdot \ldots \cdot s_N!}{k!} \delta_{t_1} \cdot \ldots \cdot \delta_{t_k} \leq$$

$$\leq \sum_{k=0}^{\infty} \frac{k! L^{\frac{1}{2}}}{k!^{\frac{1}{2}+\frac{\delta}{2}}} \cdot (\sum_{t=1}^{N} \delta_t)^k < +\infty.$$

Consequently, the series

$$\tilde{\varphi}(\lambda) = \sum_{k=0}^{\infty} (-1)^k \cdot k! (\sum_{t_1 \ldots t_k=1}^{N} \frac{\partial^k \varphi(0)}{\partial \lambda_{t_1} \ldots \partial \lambda_{t_k}} \frac{\lambda_{t_1} \cdot \ldots \cdot \lambda_{t_k}}{s_1! \cdot \ldots \cdot s_N!}$$

defines an analytical function in C^N. Due to (3.5.4)

$$E\alpha\beta = \frac{1}{(2\pi i)^N} \cdot \int_{\Gamma_R} \frac{\tilde{\varphi}(\frac{1}{\lambda_1}, \ldots, \frac{1}{\lambda_N})}{\lambda_1 \cdot \ldots \cdot \lambda_N} \cdot \psi(\lambda_1, \ldots, \lambda_N) d\lambda. \tag{3.5.5}$$

Consider $\beta \in L_2^-$ such that the series

$$\psi(\lambda) = \sum_{k=0}^{\infty} i^k \sum_{t_1 \ldots t_k=1}^{N} d_{t_1 \ldots t_k} \lambda_{t_1} \cdot \ldots \cdot \lambda_{t_k} \tag{3.5.6}$$

converges in a certain neighborhood of 0 in C^N. Then, for sufficiently small R, the action $< \alpha, \beta >$ can be obtained by (3.5.5).

Definition 3.5.1. The generalized functional $\beta \in L_2^-$ is called regular (denoting $\beta \in \mathcal{P}$) if the series (3.5.6) converges in a certain neighbourhood of 0 in C^N and

$$\psi(\lambda) = \frac{\gamma(\lambda)}{P(\lambda)}, \tag{3.5.7}$$

where P is a polynomial from λ, γ is $\mathcal{T}$-transformation of some random variables from L_2 such that

$$\sup_{k \geq 0} |\gamma^{(k)}(0)| < +\infty.$$

Every regular generalized functional can be considered as a usual random variable. Indeed let $\beta \in \mathcal{P}$, and γ, P be elements of (3.5.7). Consider, for $\epsilon > 0$, Hermander Stairs for the polynomial P $\mathcal{H}_p$ in C^N [30], that is such a system of surfaces that

1) $\inf_{\mathcal{H}_p} |P(\lambda)| > 0$

2) $\mathcal{H}_p \subset \{\lambda : \ |\text{Im } \lambda_1| + \ldots + |\text{Im } \lambda_N| < \epsilon\}$,

3) for every analytical function h in C^N which satisfies the conditions

$$\int_{R^N} |h||d\lambda| < +\infty,$$

$$\int_{\mathcal{H}_p} |h||d\lambda| < +\infty,$$

$$\lim_{R\to+\infty} R \max\{|h(\lambda)| : |\lambda_1| = \ldots = |\lambda_N| = R, |\text{Im } \lambda_1| + \ldots + |\text{Im } \lambda_N| < \epsilon\} = 0,$$

the following equality holds

$$\int_{\mathcal{H}_p} h(\lambda)(d\lambda) = \int_{R^N} h(\lambda)d\lambda.$$

Due to the condition on γ,

$$\int_{\mathcal{H}_p} \left|\exp(-\frac{1}{2}\sum_{j=1}^{N}\lambda_j^2)\cdot\frac{\gamma(\lambda)}{P(\lambda)}\cdot\exp\left(-i\sum_{j=1}^{N}x_j\lambda_j\right)\right||d\lambda| < +\infty.$$

So the functional β as a random variable $f(\xi_1,\ldots,\xi_N)$ can be considered where

$$\forall\ (x_1,\ldots,x_N) \in R^N:$$

$$f(x_1,\ldots,x_N) = \frac{1}{\sqrt{2\pi}^N}\cdot\int_{\mathcal{H}_p}\exp(-\frac{1}{2}\sum_{j=1}^{N}\lambda_j^2)\cdot\frac{\gamma(\lambda)}{P(\lambda)}\cdot$$

$$\cdot\exp(-i\sum_{j=1}^{N}x_j\lambda_j)d\lambda\cdot\exp(\frac{1}{2}\sum_{j=1}^{N}x_j^2). \tag{3.5.8}$$

The function $f\cdot\exp(-\frac{1}{2}\sum_{j=1}^{N}x_j^2)$ is a regular generalized function with the Fourier transform $\frac{\gamma}{P}$ [30]. Hence elements of $\mathcal{P}$ are usual random variables and their action on elements of L_2^+ is described by (3.5.5). Note that, for the random variable η from $\mathcal{P}\cap L_2$, formula (3.5.8) gives an f such that $\eta = f(\xi_1,\ldots,\xi_N)$.

3.6 Generalized Gaussian functionals of the second kind and the Fourier transform

Let H_+ be a real separable Hilbert space which is tightly embedded in H, with the Hilbert-Schmidt embedding $j : H_+ \to H$ such that $\sigma^2(j) < \frac{1}{2}$. Denote $H_- = H_+^*$. Then $H_+ \subset H \subset H_-$. As in the previous section the space L_2 can be considered as

$$L_{2,+} := \oplus\sum_{k=0}^{\infty}\sqrt{k!}H_{+,k},$$

$$L_{2,-} := \oplus\sum_{k=0}^{\infty}\sqrt{k!}H_{-,k}.$$

These spaces are called the spaces of the test and generalized functionals from ξ of the second kind respectively. The elements of $L_{2,+}(L_{2,-})$ will be indicated henceforth by $+(-)$. In this section only functionals of the second kind will be considered, so the term 'the second kind' will be omitted. As in the previous section

$$L_{2,+} \subset L_2 \subset L_{2,-} = L_{2,+}^*$$

and, for $\alpha_+ \in L_{2,+}$, $\beta_- \in L_{2,-}$, an action

$$< \alpha_+, \beta_- > := \sum_{k=0}^{\infty} k! < A_k, B_k >_k,$$

where

$$\alpha_+ = \oplus \sum_{k=0}^{\infty} \sqrt{k!} A_k, \ \beta_- = \oplus \sum_{k=0}^{\infty} \sqrt{k!} B_k$$

and $\langle A_k, B_k \rangle_k$ is an action of $B_k \in H_{-,k} = H_{+,k}^*$ on $A_k \in H_{+,k}$. Consider the following example.

Example 3.6.1. Denote by $\tilde{L}_{2,+}$ the complexification of $L_{2,+}$. Then, for $\varphi \in H_+$, the random variable $\exp i(\varphi, \xi) \in L_{2,+}$ and for $\varphi \in H_+$ the random variable $\exp i(\varphi, \xi) \in \tilde{L}_{2,+}$. Indeed

$$\exp i(\varphi, \xi) = \exp\left\{-\frac{\|\varphi\|^2}{2}\right\} \cdot \exp\left\{\frac{\|\varphi\|^2}{2} + \sqrt{2}\frac{i\|\varphi\|^2}{\sqrt{2}} \cdot \left(\frac{\varphi}{\|\varphi\|}, \xi\right)\right\} =$$

$$= \exp\left\{-\frac{\|\varphi\|^2}{2}\right\} \cdot \sum_{k=0}^{\infty} \frac{i^k}{k!} \cdot \|\varphi\|^k \cdot \varphi_1^{\otimes k}(\xi, \ldots, \xi).$$

Here, $\varphi_1 = \frac{\varphi}{\|\varphi\|} \in H_+$, $\varphi_1^{\otimes k}$ is the kth tensor power of φ_1 and

$$\forall\, k \geq 1: \ \|\varphi_1^{\otimes k}\|_{+,k}^2 = \|\varphi_1\|_+^{2k} = \frac{1}{\|\varphi_1\|}^{2k} \cdot \|\varphi_1\|_+^{2k}.$$

So

$$\exp i(\varphi; \xi) \in \tilde{L}_{2,+}.$$

To describe the elements of $L_{2,-}$ use the $\mathcal{T}$-transform. Define for $\alpha_- \in L_{2,-}$ and $\varphi \in H_+$

$$\mathcal{T}(\alpha_-)(\varphi) := \exp\left\{\frac{\|\varphi\|^2}{2}\right\} \cdot \langle (\varphi, \xi), \alpha_- \rangle.$$

It can be easily checked that for

$$\alpha_- = \oplus \sum_{k=0}^{\infty} \sqrt{k!} A_k,$$

$$\mathcal{T}(\alpha_-)(\varphi) = \sum_{k=0}^{\infty} i^k A_k(\varphi, \ldots, \varphi),$$

where $A_k(\varphi, \ldots, \varphi) = \langle \varphi^{\otimes k}, A_k \rangle_k$. Since

$$\sum_{k=0}^{\infty} k! \|A_k\|_{k,-}^2 < +\infty,$$

then

$$\exists\, C > 0 : \forall\, k \geq 0 : \|A_k\|_{k,-} < \frac{C}{\sqrt{k!}}.$$

Consequently the function $\mathcal{T}(\alpha_-)$ is analytical on H_+. Hence the $\mathcal{T}$ transform is an injection as similar to the Fourier transform. Now consider an example of generalized functionals. Suppose that there are two probability spaces $(\Omega_1, \mathcal{F}_1, P_1)$, $(\Omega_2, \mathcal{F}_2, P_2)$ and two generalized random Gaussian elements in H ξ_1, ξ_2 on these spaces which satisfy all conditions for ξ. The functionals from $\xi_j, j = 1, 2$ will be indicated by j.

Example 3.6.2. Define the inner product (ξ_1, ξ_2) as a generalized functional from ξ_2 which depends on ξ_1. Let $\{\varphi_k;\ k \geq 1\}$ be an orthonormal basis in H. It is natural to define (ξ_1, ξ_2) with the help of the sequence

$$S_n = \sum_{k=1}^{n} (\varphi_k, \xi_1)(\varphi_k, \xi_2),\ n \geq 1.$$

Note that, under $\omega_2 \in \Omega_2$ being fixed, $\{S_n;\ n \geq 1\}$ is the sequence of partial sums of the series from the independent identically distributed Gaussian random variables with zero mean variance 1 and coefficients $\{(\varphi_k, \xi_2)(\omega_2);\ k \geq 1\}$. Since

$$P_2\{\sup_{k \geq 1} |(\varphi_k, \xi_2)(\omega_2)| < +\infty\} = 0,$$

then $\{S_n;\ n \geq 1\}$ does not converge on Ω_1, in any of usual senses under $\omega_2 \in \Omega_2$ being fixed with P_2-probability 1. Consider the $\mathcal{T}_1$-transform from $\{S_n;\ n \geq 1\}$:

$$\forall\, \varphi \in H_+ :\ \mathcal{T}_1(S_n)(\varphi) = i \exp\left\{-\frac{\|\varphi\|^2}{2}\right\} \cdot \sum_{k=1}^{n} (\varphi_k, \xi_2) \cdot (\varphi_k, \varphi).$$

The sequence $\{\mathcal{T}_1(S_n)(\varphi);\ n \geq 1\}$ converges in the square-mean with respect to P_2, to the random variable

$$\gamma_2 = i \exp\left\{-\frac{\|\varphi\|^2}{2}\right\} \cdot (\varphi, \xi_2).$$

Since the embedding $H \subset H_-$ is Hilbert-Schmidt and $\varphi \in H_+$, then (φ, ξ_2) can be regarded as an action of the random element $\xi_2 \in H_-$ on the element $\varphi \in H_+$. So γ_2 is the $\mathcal{T}_1$-transform of a certain element of $L_{2,-,1}$. This element will be considered as (ξ_1, ξ_2).

Example 3.6.3. Define $\exp i(\xi_1, \xi_2)$. As in the previous example, consider the sequence

$$\Phi_n = \exp\{i \sum_{k=1}^{n} (\varphi_k, \xi_1)(\varphi_k, \xi_2)\}\ ,\ n \geq 1.$$

For every $n \geq 1$

$$M_1\Phi_n = \exp\{-\frac{1}{2}\sum_{k=1}^{n}(\varphi_k, \xi_2)^2\}.$$

Hence

$$T_1(\frac{\Phi_n}{M_1\Phi_n})(\varphi) = \exp\{\sum_{k=1}^{n}(\varphi_k, \varphi)(\varphi_k, \xi_2)\} \ , \ n \geq 1, \ \varphi \in H_+.$$

Consequently, for every $\varphi \in H_+$ the sequence $\{T_1(\frac{\Phi_n}{M_1\Phi_n})(\varphi); n \geq 1\}$ converges in the square-mean with respect to P_2 (sf. [31]) to the random variable

$$\exp(\varphi, \xi_2) = \sum_{k=0}^{\infty} i^n \frac{(-i)^n}{n!}(\xi_2, \varphi)^n,$$

which is P_2 - a.s. T_1 - transform of an element from $L_{2,-,1}$.

The operations of the stochastic differentiation and integration can be defined for generalized random functionals.

Definition 3.6.1. For $\alpha_- \in L_2$ with the expansion

$$\alpha_- = \oplus \sum_{k=0}^{\infty} \sqrt{k!}A_k \ ,$$

and $\varphi \in H_+$, the partial derivative $\partial_\varphi \alpha_-$ is the sum

$$\partial_\varphi \alpha_- := \oplus \sum_{k=1}^{\infty} \sqrt{(k-1)!}kA_k(\varphi, \ldots) \ ,$$

if this sum belongs to $L_{2,-}$.

For example, in the case $\alpha_- \in W^1$ the partial derivative $\partial_\varphi \alpha_-$ exists for every $\varphi \in H_+$ and

$$\partial_\varphi \alpha_- = (D\alpha_-, \varphi).$$

If $\varphi \in H_+$ is fixed, then ∂_φ is a closed linear operator in $L_{2,-}$ and its domain of definition is tight in $L_{2,-}$. The adjoint operator ∂_φ^* has the following appearance

$$\partial_\varphi^* = \oplus \sum_{k=o}^{\infty} \sqrt{(k+1)!}\Lambda\varphi \otimes A_k \ ,$$

where Λ is a symmetrization operator and φ in the right part, is considered as an element of H_-. For example, if the random element $x \in W^1(H)$, the extended stochastic integral $I(x)$ can be obtained in such a way:

$$I(x) = \sum_{k=1}^{\infty} \partial_{\varphi_k}^*(x, \varphi_k) \ ,$$

where $\{\varphi_k; k \geq 1\} \subset H_+$ is an orthonormal basis in H.

It follows from theorem 3.4.1 , that, for $\alpha \in W^1$,

$$\partial_\varphi \alpha + \partial_\varphi^* \alpha = (\varphi, \xi)\alpha \ .$$

Define now the generalized Fourier transform for the random test variables. For $\alpha_{+,1}$

$$\mathcal{F}(\alpha_{+,1}) := \langle \exp i(\xi_1,\ \xi_2), \alpha_{+,1} \rangle.$$

This definition is proper due to the statement of example 3.6.3.

Lemma 3.6.1. $\mathcal{F}$ is a continuous linear operator from $L_{2,+,1}$ to $\tilde{L}_{2,2}$.

Proof. Let

$$\alpha_{+,1} = \oplus \sum_{n=0}^{\infty} \sqrt{n!} A_n.$$

Then, due to example 3.6.3 ,

$$\mathcal{F}(\alpha_{+,1}) = \sum_{n=0}^{\infty} (-i)^n A_n[\xi_2, \ldots, \xi_2].$$

where $A_n[\xi, \ldots, \xi]$ is an action of the n-linear form $A_n \in H_{+,n} = (H_-^*)^n$ on the random element $\xi_2 \in H_-$. Under this

$$M_2 |\mathcal{F}(\alpha_{+,1})|^2 = M_2 \sum_{mn=0}^{\infty} (-i)^{(m-n)} A_m[\xi_2, \ldots, \xi_2].$$

$$A_n[\xi_2, \ldots, \xi_2] \leq \sum_{mn=0}^{\infty} |A_m|_{m,+} \cdot A_n|_{n,+} \cdot M_2 |\xi_2|_-^{m+n} \leq$$

$$\leq \|\alpha_{+,1}\|_+^2 \cdot \sum_{mn=0}^{\infty} \frac{1}{\sqrt{m!}} \frac{1}{\sqrt{n!}} \cdot \sqrt{2} \sqrt{(m+n)!} (\sigma^2(j))^{(\frac{m+n}{2})} \leq$$

$$\leq \|\alpha_{+,1}\|_+^2 \cdot \sum_{k=0}^{\infty} (k+1) 2^{\frac{k}{2}} \cdot (\sigma^2(j))^{(\frac{k}{2})} = C \cdot \|\alpha_{+,1}\|_+^2.$$

Here $\| \cdot \|_+$ is the norm in $L_{2,+,1}$ and

$$C = \sum_{k=0}^{\infty} (k+1) 2^{\frac{k}{2}} (\sigma^2(j))^{\frac{k}{2}} < +\infty$$

because $\sigma^2(j) < \frac{1}{2}$. Lemma is proved.

The value $\mathcal{F}(\alpha_{+,1})$ can be obtained in such a way:

$$\mathcal{F}(\alpha_{+,1}) = F(\xi_2), F \in C(H_-).$$

Here

$$F(\varphi) = \mathcal{E}(\beta_{+,1})(\varphi),\ \varphi \in H_+ ,$$

where

$$\beta_{+,1} = \oplus \sum_{n=0}^{\infty} (-1)^n \sqrt{n!} A_n ,$$

and $\mathcal{E}$ is the Fourier-Wiener transform.

Lemma 3.6.2. Let, for $\alpha_{+,1} \in L_{2,+,1}$, $\varphi \in H_+$, there exist $\partial_\varphi \alpha_{+,1}$, $\partial_\varphi^* \alpha_{+,1} \in L_{2,+,1}$. Then $(\varphi, \xi_1) \cdot \alpha_{+,1} \in L_{2,+,1}$ and

$$\mathcal{F}((\varphi, \xi_1) \cdot \alpha_{+,1}) = -i \partial_\varphi^* \cdot \mathcal{F}(\alpha_{+,1}) ,$$

$$\mathfrak{F}(\partial_\varphi^* \cdot \alpha_{+,1}) = -i(\varphi, \xi_2) \cdot \mathfrak{F}(\alpha_{+,1}) \ .$$

Proof. The inclusion $(\varphi, \xi_1) \cdot \alpha_{+,1} \in L_{2,+,1}$ follows from the equality

$$(\varphi, \xi_1) \cdot \alpha_{+,1} = \partial_\varphi \alpha_{+,1} + \partial_\varphi^* \alpha_{+,1} \ .$$

Since $\mathfrak{F}$ is continuous and the operator of multiplication by $(\varphi, \ \xi_2)$ is closed as well as ∂_φ and ∂_φ^*, then it is sufficient to check the statement for the case

$$\alpha_{+,1} = A_n, \ A_n \in H_{+,n}, \ n \geq 0 \ .$$

From the definitions

$$\partial_\varphi \alpha_{+,1} = nA_n(\varphi, \xi_1, \ldots, \xi_1),$$
$$\partial_\varphi^* \alpha_{+,1} = \Lambda\varphi \otimes A_n(\xi_1, \ldots, \xi_1).$$

Hence

$$\mathfrak{F}(\partial_\varphi \alpha_{+,1}) = (-i)^{n-1} nA_n[\varphi, \xi_2, \ldots, \xi_2],$$
$$\mathfrak{F}(\partial_\varphi^* \alpha_{+,1}) = (-i)^{n+1} \Lambda\varphi \otimes A_n[\xi_2, \ldots, \xi_2] =$$
$$(-i)^{n+1}(\varphi, \xi_2) A_n[\xi_2, \ldots, \xi_2] = (-i)(\varphi, \xi_2) \cdot \mathfrak{F}(\alpha_{+,1}) \ .$$

Consequently

$$\mathfrak{F}((\varphi, \xi_1) = \mathfrak{F}(\partial_\varphi \alpha_{+,1} + \partial_\varphi^* \alpha_{+,1}) =$$
$$= (-i)^{n-1} nA_n[\varphi, \xi_2, \ldots, \xi_2] + (-i)^{n+1}(\varphi, \xi_2) A_n[\xi_2, \ldots, \xi_2] =$$
$$= (-i)((\varphi, \xi_2)\mathfrak{F}(\alpha_{+,1}) - \partial_\varphi \mathfrak{F}(\alpha_{+,1})) =$$
$$= (-i)\partial_\varphi^* \mathfrak{F}(\alpha_{+,1}) \ .$$

Lemma is proved.

The transform $\mathfrak{F}$ can be defined for the random elements from $L_2(\Omega_1, P_1, X)$, which belong to $X \otimes L_{2,+,1}$. For $x \in X \otimes L_{2,+,1}$, which has the expansion

$$x = \sum_{i=1}^{\infty} x_i \alpha_{i,+,1},$$

define

$$\mathfrak{F}(x) := \sum_{i=1}^{\infty} x_i \mathfrak{F}(\alpha_{i,+,1}).$$

It follows from lemma 3.6.3 that $\mathfrak{F}$ is a continuous linear operator from $X \otimes L_{2,+,1}$ to $L_2(\Omega_2, P_2, \tilde{X})$, where $\tilde{X}$ is a completion of X.

Let A_1 be a random operator in $X and$ $\{e_k; k \geq 1\}$ be an orthonormal basis in X. Then

$$\forall x \in X: \ A_1(x) = \sum_{k=1}^{\infty} \{\sum_{l=1}^{\infty} [A_1(e_l), e_k] \cdot [x, e_l]\} e_k,$$

where both series converge in the square-mean. Suppose that

$$\forall k, l \geq 1: \ [A_1 e_l; e_k] \in L_{2,+,1} \ .$$

Define the random operator $\mathcal{F}(A_1)$ in such a way:

$$\forall x \in X: \quad \mathcal{F}(A_1)(x) := \sum_{k=1}^{\infty}\{\sum_{l=1}^{\infty}\mathcal{F}([A_1(e_l), e_k]) \cdot [x, e_l]\}e_k,$$

if both series converge in the square-mean and the result continuously depends on x with the same meaning.

Theorem 3.6.1. Let A_1 be such a random operator in X that there exists $\mathcal{F}(A_1)$. Suppose that, for the random element $x \in W^1(X) \cap X \otimes L_{2,+,1}$ and for a certain orthonormal basis $\{e_k; k \geq 1\}$ in X, the series

$$A_1 \circ x := \sum_{k=1}^{\infty}[x, e_k] \cdot A_1(e_k)$$

converges in the square-mean to an element of $L_{2,+,1}$. Then

$$\mathcal{F}(A_1 \circ x) = \mathcal{F}(A_1)(\mathcal{F}(x)) \ ,$$

where $\mathcal{F}$, in the right part, is defined as in section 3.2.

The proof of the theorem is similar to the proof of lemma 3.6.3.

Example 3.6.4. Let $H = X = L_2([0;1])$ and the generalized random elements ξ_1, ξ_2 be defined with the help of the corresponding Wiener processes w_1, w_2:

$$\forall \varphi \in H : (\varphi, \xi_j) := \int_0^1 \varphi(\tau)dw_j(\tau).$$

Define H_+ as Sobolev space $W_2^1([0;1])$ and $H_- = H_+ = W_2^{-1}([0;1])$. For every $\varphi \in W_2^1([0;1])$,

$$\|\varphi\|_+^2 := 4(\int_0^1 \varphi^2(\tau)d\tau + \int_0^1 (\varphi'(\tau))^2 d\tau) \ .$$

Now the space $L_{2,+,1}$ consists of such random variables $\alpha_{+,1}$ for which, in the expansion

$$\alpha_{+,1} = \sum_{k=0}^{\infty}\int_{\Delta_k(1)} a_k(\tau_1, \ldots, \tau_k)dw(\tau_1)\ldots dw(\tau_k),$$

all the kernels $\{a_k; k \geq 1\}$ have generalized derivatives of the first order with respect to convergence in the square-mean. Consider the random functional

$$\forall \varphi \in L_2([0;1]): \ F_1(\varphi) = \int_0^1 \varphi(\tau)dw_1(\tau).$$

Then $\mathcal{F}(F_1)$ is correctly determined and

$$\forall \varphi \in L_2([0;1]): \ \mathcal{F}_1(\varphi) = -i\int_0^1 \varphi(\tau)dw_2(\tau).$$

Indeed if $\{f_n; n \geq 1\}$ is an orthonormal basis of continuously differentiable functions in $L_2([0;1])$, then

$$\forall n \geq 1 : \int_0^1 f_n(\tau)dw_1(\tau) \in L_{2,+,1} \ ,$$

$$\mathfrak{F}(\int_0^1 f_n(\tau)dw_1(\tau)) = -i\langle f_n,\ \dot{w}_2\rangle,$$

where the right part is an action of $\dot{w}_2$ as an element of $W_2^{-1}([0;1])$ on $f_n \in W_2^1([0;1])$. It can be easily checked, that

$$\langle f_n,\ \dot{w}_2\rangle = \int_0^1 f_n(\tau)dw_2(\tau)\ .$$

So $\forall n \geq 1$:

$$\mathfrak{F}(\int_0^1 f_n(\tau)dw_1(\tau)) = -i\int_0^1 f_n(\tau)dw_2(\tau)\ .$$

Consequently it can be defined that $\forall\varphi \in L_2([0;1])$:

$$\mathfrak{F}(F_1)(\varphi) = \sum_{n=1}^{\infty}(-i)\int_0^1 f_n(\tau)dw_2(\tau)\cdot\int_0^1 f_n(\tau)\varphi(\tau)d\tau = -i\int_0^1 \varphi(\tau)dw_2(\tau)\ .$$

Consider now the random function $x \in W^1(L_2([0;1]))$, which belongs to $L_2([0;1])\otimes L_{2,+,1}$, such that the series

$$\sum_{n=1}^{\infty}\int_0^1 f_n(\tau)dw_1(\tau)\cdot\int_0^1 f_n(\tau)x(\tau)d\tau$$

converges to the random variable from $L_{2,+,1}$. Then

$$\mathfrak{F}(\sum_{n=1}^{\infty}\int_0^1 f_n(\tau)dw_1(\tau)\cdot\int_0^1 f_n(\tau)x(\tau)d\tau) = -i\int_0^1 \mathfrak{F}(x)(\tau)dw_2(\tau)\ .$$

Hence, in this case, $\mathfrak{F}$ transforms the symmetric stochastic integral into the extended stochastic integral.

Chapter 4

Stochastic analysis and quantum mechanics

4.1 Systems with changeable numbers of particles

The aim of this chapter is to consider some mathematical models which have a "physical" origin and are connected with the notions of previous chapters. This section is devoted to systems of particles which satisfy the Bose-Einstein statistics. The analogy between stochastic calculus and the description of such systems has been mentioned by many authors (cf. [32]). So this section does not contain any new statements. All physical facts which are considered can be found in [33].

Consider the system with a changeable number of particles. To describe the states of such a system use the Hilbert space $\mathcal{H}$ which corresponds to the set of states of one particle. In physics the space $\mathcal{H}$ must be complex, but suppose that $\mathcal{H}$ is a real space to preserve the analogy with the space of white noise. Let $(\cdot,\cdot)$ and $\|\cdot\|$ be an inner product and a norm in $\mathcal{H}$ respectively. Then the states of the system of n particles are described by the elements of the n-th tensor power of $\mathcal{H}$. Since it has been supposed that the particles are equal then the system from n particles is described by the symmetric part of $\mathcal{H}^{\otimes n}$. Denote this symmetric part by $\mathcal{H}_n$. Note that $\mathcal{H}_n$ is the space of all symmetric Hilbert-Schmidt n-linear form on $\mathcal{H}$. The states of the system with a changeable number of particles are the elements of such an infinite sum as

$$\mathcal{H}_B := \oplus \sum_{n=0}^{\infty} \mathcal{H}_n,$$

where $\mathcal{H}_0 := R$. For every element $(A_n;\ n \geq 0)$ of $\mathcal{H}_B$ an analytical function on $\mathcal{H}$ can be considered thus,

$$\forall\, \varphi \in \mathcal{H} : \sum_{n=0}^{\infty} \frac{1}{\sqrt{n!}} A_n(\varphi, \ldots, \varphi).$$

Define, for the fixed $\varphi \in \mathcal{H}$, the operators of the creation and the annihilation of the particles in state φ:

$$\forall\, n \geq 0, A_n \in \mathcal{H}_n :$$

$$a(\varphi)(A_n) := \sqrt{n} A_n(\varphi, \ldots).$$

For the sum $\oplus \sum_{n=0}^{N} A_n$, an action of $a(\varphi)$ is defined by linearity. The closure of $a(\varphi)$ in $\mathcal{H}_B$ is called by the annihilation operator in state φ. Consider some properties of $a(\varphi)$.

1. $a(\varphi)$ continuously transforms $\mathcal{H}_n$ into $\mathcal{H}_{n-1}$, for $n \geq 1$.
2. For $\varphi, \psi \in \mathcal{H}$ the operators $a(\varphi)$ and $a(\psi)$ are commuting on $\mathcal{H}_n$, $n \geq 0$.

The adjoint operator $a(\varphi)^*$ is called by the creation operator in state φ. It can be easily checked that, for every $A_n \in \mathcal{H}_n$,

$$a(\varphi)^*(A_n) = \sqrt{n+1} \cdot \Lambda \varphi \otimes A_n,$$

where Λ is a symmetrization operator. The creation operators have the following properties:

1. For every $\varphi \in \mathcal{H}$, $n \geq 0$ $a(\varphi)^*$ is a bounded linear operator from $\mathcal{H}_n$ to $\mathcal{H}_{n+1}$;
2. For every $\varphi, \psi \in \mathcal{H}$ the operators $a(\varphi)^*$ and $a(\psi)^*$ are commuting on $\mathcal{H}_n$, $n \geq 0$.

The operators of creation and annihilation are connected with one another by an equality:

$$\forall\, \varphi, \psi \in \mathcal{H} :$$

$$a(\varphi)a(\psi)^* - a(\psi)^* a(\varphi) = (\psi, \varphi) \cdot E,$$

where E is the identity operator. All elements of $\mathcal{H}_B$ can be constructed with the help of creation operators. To do this, the creation operator is built not only for $\varphi \in \mathcal{H}$, but for every $A_n \in \mathcal{H}_n$, where $n \geq 2$.

If $A_n \in \mathcal{H}_n$, $n \geq 2$ then

$$A_n = \sum_{t_1 \ldots t_n = 1}^{\infty} \alpha_{t_1 \ldots t_n} \cdot \varphi_{t_1} \otimes \cdots \otimes \varphi_{t_n},$$

where $\{\varphi_t;\ t \geq 1\}$ is an orthonormal basis in $\mathcal{H}$, the values $\{\alpha_{t_1 \ldots t_n}\}$ do not depend on the permutation of $t_1 \ldots t_n$ and

$$\sum_{t_1 \ldots t_n}^{\infty} \alpha_{t_1 \ldots t_n}^2 = \|A_n\|_n^2.$$

Define

$$a(A_n)^* := \frac{1}{\sqrt{n!}} \sum_{t_1 \ldots t_n}^{\infty} \alpha_{t_1 \ldots t_n} \cdot a(\varphi_{t_1})^* \cdot \ldots \cdot a(\varphi_{t_n})^*.$$

Such an operator is defined on the dense subset of $\mathcal{H}_B$ and

$$A_n = a(A_n)^*(\eta_0),$$

where $\eta_0 = 1$ (more carefully $\eta_0 = (1, 0, \ldots, 0, \ldots)$).

Consequently

$$\forall\, \eta = (A_n;\ n \geq 0) \in \mathcal{H}_B :$$

$$\eta = \sum_{n=0}^{\infty} a(A_n)^*(\eta_0).$$

Where, for $n = 0$, $a(A_0)^*$ is a multiplication operator on the value of A_0.

Define one more operator such that

$$N := \sum_{t=1}^{\infty} a(\varphi_t)^* a(\varphi_t).$$

The operator N is called the number of particles operator because:

$$\forall\, n \geq 0 \;\forall\, A_n \in \mathcal{H}_n :$$

$$N(A_n) = n \cdot A_n.$$

The closure of N is a self-adjoint non-negative operator in $\mathcal{H}_B$ and

$$\forall\, \varphi, \psi \in \mathcal{H} :$$

$$\mathcal{D}(N^{\frac{1}{2}}) \subset \mathcal{D}(a(\varphi)) \cap \mathcal{D}(a(\psi)^*).$$

Consider the situation of section 1.1, when the generalized random element ξ 'takes values' in $\mathcal{H}$. Then space $L_2(\Omega, \mathcal{F}, P)$ can be identified with $\mathcal{H}_B$ using the map under which the random variable $\sum_{n=0}^{\infty} A_n(\xi, \ldots, \xi)$ corresponds to the sequence $(\sqrt{n!}A_n;\ n \geq 0)$. To find 'the probability sense' of the creation and annihilation operators, suppose that $(A_n; n \geq 0) \in \mathcal{D}(N)$. Then

$$\forall\, \varphi \in \mathcal{H} :$$

$$a(\varphi)(A_n;\ n \geq 0) = (\sqrt{n}A_n(\varphi, \ldots); n \geq 0),$$

$$a(\varphi)^*(A_n;\ n \geq 0) = (\sqrt{n+1}\Lambda\varphi \otimes A_n; n \geq 0).$$

The random variable $\alpha = \sum_{n=0}^{\infty} \frac{1}{\sqrt{n!}} A_n(\xi, \ldots, \xi) = \sum_{n=0}^{\infty} \tilde{A}_n(\xi, \ldots, \xi)$ corresponds to the sequence $(A_n;\ n \geq 0)$, since the random variables

$$\alpha_1 = \sum_{n=0}^{\infty} n\tilde{A}_n(\varphi, \xi, \ldots, \xi),$$

$$\alpha_2 = \sum_{n=0}^{\infty} \Lambda\varphi \otimes \tilde{A}_n(\xi, \ldots, \xi)$$

correspond to $a(\varphi)(A_n;\ n \geq 0)$ and $a(\varphi)^*(A_n;\ n \geq 0)$. Note that $\alpha_1 = \partial_\varphi \alpha$, $\alpha_2 = \partial_\varphi^* \alpha$. Let $\{\varphi_k;\ k \geq 1\}$ be an orthonormal basis in $\mathcal{H}$ and $x \in W^1(\mathcal{H})$. Then

$$I(x) = \sum_{k=0}^{\infty} \partial_{\varphi_k}^*(x; \varphi_k).$$

In a similar way it is easy to check that the 'physical' operator N corresponds to the operator $I(D)$. So the random elements from $L_2(\Omega, P, \mathcal{H})$ can be identified with the states of the countable family of systems with changeable numbers of particles. The action of random operators can be identified with the action of certain creation operators.

4.2 The random statistics

Let $\{x_n; n \geq 1\}$ be the sequence of independent identically distributed random variables with the distribution function F. The limit theorem for symmetric statistics from $\{x_n; n \geq 1\}$ and relationship between the random statistics and the extended stochastic integral will be considered in this subchapter . Denote by H the subspace of $L_2(R, dF)$, which consists of such functions f for which

$$\int_R f(t)dF(t) = Ef(x_1) = 0.$$

Build $H_n, n \geq 0$ as before. For every $f_n \in H_n$

$$f(x_1, \ldots, x_n) = f(x_{\sigma(1)}, \ldots, x_{\sigma(n)}),$$

$$E\{f(x_1, \ldots, x_n)/x_1, \ldots, x_{n-1}\} = 0,$$

where σ is an arbitrary permutation of $\{1, \ldots, n\}$. Let ξ be a generalized Gaussian random element in H with zero mean and identity variation. Consider an arbitrary functional η from ξ with finite second moment:

$$\eta = \sum_{k=0}^{\infty} f_k(\xi, \ldots, \xi),$$

where

$$f_k(\xi, \ldots, \xi)$$

is the value of the k-linear Hilbert-Schmidt form which is denoted by f on ξ. Build the following sequence of statistics from $\{x_n; n \geq 1\}$:

$$S_N = \sum_{k=0}^{N} \frac{1}{N^{\frac{k}{2}}} \cdot \sum_{\{t_1 \ldots t_k\}} f_k(x_{t_1}, \ldots, x_{t_k}), N \geq 1.$$

In the second sum, the sets $\{t_1, \ldots, t_k\}$ do not contain equal indices and $t_1, \ldots, t_k \leq N$. The following statement is true.

Theorem 4.2.1. The sequence $\{S_N; N \geq 1\}$ converges to η in the distribution under $N \to \infty$.

Proof. Let $\{\varphi_i; i \geq 1\}$ be an orthonormal basis in H such that

$$\forall\, i \geq 1: \quad \text{essup}\, |\varphi_i| < +\infty.$$

Such a basis exists because the set $\{\cos \lambda x, \sin \lambda x,\ \lambda \in R\}$ is tight in $L_2(R, dF)$. Let $A_{l_1} \in H_{l_1}, \ldots, A_{l_k} \in H_{l_k}$ be the form which depends only on the first n coordinates of the basis $\{\varphi_i; i \geq 1\}$. Denote by $a_{l_1}, \ldots, a_{l_k}$ the kernels of $A_{l_1}, \ldots, A_{l_k}$. Prove that the joint distribution of random variables

$$S_N(a_{l_i}) = \frac{1}{N^{\frac{l_i}{2}}} \cdot \sum_{t_1, \ldots, t_{l_i}} a_{l_i}(x_{t_1}, \ldots, x_{t_{l_i}}), i = 1, \ldots, k \tag{4.2.2}$$

converges with the joint distribution of $A_{l_1}(\xi,\dots,\xi),\dots,A_{l_k}(\xi,\dots,\xi)$. Use the method of mathematical induction for $l=\max\{l_i; i=1,\dots,k\}$. Under $l=1$, the random variables $S_N(a_1^1)\dots S_N(a_1^k)$ are such that

$$S_N(a_1^i)=\frac{1}{N^{\frac{1}{2}}}\cdot\sum_{t_1=1}^{N}\sum_{j=1}^{n}\alpha_{ij}\varphi_j(x_{t_1}), i=1,\dots,k,$$

and the statement follows from the central limit theorem in R^k. Suppose that the statement is true under $l=m\geq 1$ for an arbitrary $A_{l_1},\dots,A_{l_k}$, $n,k\geq 1$. Consider the set $A_{l_1},\dots,A_{l_k},B^1_{m+1},\dots,B^r_{m+1}$ where $l_i\leq m, i=1,\dots,k$ and all forms depend only on the first n coordinate of the basis $\{\varphi_i; i\geq 1\}$. Let b_1 be the kernel of B^1_{m+1}. The random variable $S_N(b^1)$ can be written in the following manner :

$$S_N(b_1)=\frac{1}{N^{\frac{m+1}{2}}}\cdot\sum_{-t_1,\dots,t_{m+1}-}b^1(x_{t_1},\dots,x_{t_{m+1}})=$$

$$=\frac{1}{N^{\frac{m+1}{2}}}\cdot\sum_{-t_1,\dots,t_{m+1}-}\sum_{i_1,\dots,i_{m+1}=1}^{n}\beta_{i_1,\dots,i_{m+1}}\varphi_{i_1}(x_{t_1})\cdot\ldots\cdot\varphi_{i_{m+1}}(x_{t_{m+1}})=$$

$$=\frac{1}{N^{\frac{m+1}{2}}}\cdot\sum_{i_1=1}^{n}\sum_{-t_1,\dots,t_{m+1}-}\sum_{i_2,\dots,i_{m+1}=1}^{n}\beta_{i_1,\dots,i_{m+1}}\varphi_{i_1}(x_{t_1})\cdot\ldots\cdot\varphi_{i_{m+1}}(x_{t_{m+1}})=$$

$$=\frac{1}{N^{\frac{m+1}{2}}}\cdot\sum_{i_1=1}^{n}\Bigg\{\sum_{t_1=1}^{N}\varphi_{i_1}(x_{t_1})\cdot\sum_{-t_2,\dots,t_{m+1}-}\sum_{i_2,\dots,i_{m+1}=1}^{n}\beta_{i_1,\dots,i_n}\varphi_{i_2}(x_{t_2})\cdot\ldots\cdot\varphi_{i_{m+1}}(x_{t_{m+1}})-$$

$$-m\sum_{-t_2,\dots,t_{m+1}-}\sum_{i_2,\dots,i_{m+1}=1}^{n}\beta_{i_1,\dots,i_n}\varphi_{i_1}(x_{t_2})\varphi_{i_2}(x_{t_2})\cdot\ldots\cdot\varphi_{i_{m+1}}(x_{t_{m+1}})\Bigg\}$$

It can be verified, by continuation of this process, that $S_N(b^1)$ is a finite sum of such random variables

$$\frac{1}{N^{\frac{r}{2}}}\sum_{t_1=1}^{N}\varphi_{i_1}(x_{t_1})\cdot\ldots\cdot\varphi_{i_r}(x_{t_1})\cdot S_N(c^1_{i_1\dots i_r}),$$

where $r=1,\dots,m+1$ and $c^1_{i_1\dots i_r}\in H_{m+1-r}$. Since the functions $\{\varphi_i;\ i\geq 1\}$ are bounded on R then, due to the law of large numbers, under $r=2$

$$\frac{1}{N}\sum_{t_1=1}^{N}\varphi_{i_1}(x_{t_1})\varphi_{i_2}(x_{t_2})\to E\varphi_{i_1}(x_{t_1})\cdot\varphi_{i_2}(x_{t_2})=(\varphi_{i_1},\varphi_{i_2})\ (\text{mod P}),$$

when $N\to\infty$. Under $r>2$

$$\frac{1}{N^{\frac{r}{2}}}\cdot\sum_{t_1=1}^{N}\varphi_{i_1}(x_{t_1})\cdot\ldots\cdot\varphi_{i_r}(x_{t_1})\to 0\quad(\text{mod }P), N\to\infty.$$

It follows, from the central limit theorem, that

$$\frac{1}{N^{\frac{1}{2}}}\cdot\sum_{t_1=1}^{N}\varphi_{i_1}(x_{t_1})\Rightarrow(\varphi_{i_1},\xi), N\to\infty.$$

To prove the statement it is enough to use the induction supposition and the following fact. If the sequences of random vectors in the finite-dimensional space $\{\eta_n^1;\ n \geq 1\}$, $\{\eta_n^2;\ n \geq 1\}$ are such that $\{\eta_n^1;\ n \geq 1\}$ converges in the distribution to η^1 and $\{\eta_n^2;\ n \geq 1\}$ converges with probability 1 to the non-random c, then $\{[\eta_n^1, \eta_n^2];\ n \geq 1\}$ converges in the distribution to $[\eta^1, c]$. The proof of the theorem can be completed with the help of the approximation of the series for η by the finite sums of the finite-dimensional forms.

Note that the similar statement is true for the random element η in $H_k, k \geq 1$, i. e. for the H_k-valued statistics. The random elements in H can be considered as random statistics.

Let $\{\varphi_i; i \geq 1\}$ be the orthonormal basis in H from theorem 4.2.1. Consider the following random operator in H,

$$\forall\, \varphi \in H:\ A(\varphi) = \sum_{i=1}^{-}(\varphi_i;\xi)(\varphi_i;\varphi)\cdot\varphi_i$$

and the random element $\eta = (\varphi_1;\xi)\varphi_1$. The random element η belongs to the domain of the definition of A and the correspondent random statistics

$$S_N(\eta) = \frac{1}{N^{\frac{1}{2}}}\cdot\sum_{t_1=1}^{N}\varphi_1(x_t)\cdot\varphi_1(\cdot),$$

$$S_N\left(A(\eta)\right) = \frac{1}{N}\cdot\sum_{t_1,t_2}\varphi_1(x_{t_1})\varphi_1(x_{t_2})\cdot\varphi_1(\cdot)$$

converge in the distribution to η and $A(\eta)$.

In conclusion, note that theorem 4.2.1 in this particular case contains the well-known statements about the convergence of U-statistics to multiple Wiener integrals [34].

4.3 Local intersection times

Let $(\mathcal{X}, \mathcal{A}, \mu)$ be measurable space with the finite measure μ. Consider $\mathcal{X}$ as a phase space of the particles. The motion of the particles is described by the homogeneous Markovian process $\{x(t); t \geq 0\}$ in $(\mathcal{X}, \mathcal{A})$ with initial distribution ν and transition probability P, which satisfies the conditions

1) $\forall\, t > 0, \forall\, x \in \mathcal{X} : \mathrm{P}(t, x, \cdot) \ll \mu$,

2) $\dfrac{dP(t,x,\cdot)}{d\mu}(y) = p(t,x,y) = p(t,y,x),\ x, y \in \mathcal{X}$.

Let $\varphi : \mathcal{X} \to R$ be a bounded measurable function. Then the value

$$L(\varphi, x) := \int_0^T \varphi(x(t))dt$$

can be considered as an energy, which has been acquired by the particle during the time interval $[0;T]$ in the potential φ. Henceforth, only such functions φ shall be considered for which

$$E\int_0^T \varphi\left(x(t)\right)dt = \int_0^T \int_{\mathcal{X}^2} \varphi(y)p(t,x,y)\nu(dx)\mu(dy)dt = 0. \qquad (4.3.1)$$

Denote by $\mathcal{M}$ the set of all bounded measurable functions which satisfy (4.3.1). For $\varphi, \psi \in \mathcal{M}$ define the inner product

$$(\varphi,\psi) := EL(\varphi,x)\cdot L(\psi,x) = E\int\int_0^T \varphi(x(t_1))\cdot\psi(x(t_2))dt_1dt_2 =$$

$$= \int\int_{-^2} \varphi(x_1)h(x_1,x_2)\psi(x_2)\mu(dx_1)\mu(dx_2),$$

where

$$\forall\, x_1,x_2 \in \mathcal{X}:\ h(x_1,x_2) = \int_{\Delta_2(T)}\int_{-} p(t_1,x,x_1)\cdot p(t_2-t_1,x_1,x_2)\nu(dx)\cdot dt_1dt_2 +$$

$$+\int_{\Delta_2(T)}\int_{-} p(t_2,x,x_1)\cdot p(t_1-t_2,x_2,x_1)\nu(dx)\cdot dt_2dt_1.$$

Let H be the Hilbert space which is obtained from $\mathcal{M}$ after the factorization and completion with respect to $(\cdot,\cdot)$. For an arbitrary $\kappa \in H\backslash\mathcal{M}$, define $L(\kappa,x)$ as a limit in $L_2(\Omega,\mathcal{F},\mathrm{P})$ of the sequence $\{L(\varphi_n,x), n\geq 1\}$ where $\varphi_n \to \kappa, n\to\infty$ in H, $\varphi_n \in \mathcal{M}, n \geq 1$. The value $L(\kappa,x)$ is independent of the choice of sequence $\{\varphi_n; n\geq 1\}$.

Consider the sequence $\{x_i; i\geq 1\}$ of an independent copy of x with common initial conditions. For $\varphi \in H$ the energy of the motion of such a system in the potential φ, during the time $[0;T]$, can be described as a weak limit of the sums

$$S_n = \frac{1}{\sqrt{n}}\cdot\sum_{i=1}^{n} L(\varphi,x_i), n\geq 1,$$

where $\{x_i; i\geq 1\}$ are the independent copies of x. Due to the central limit theorem the sequence $\{S_n;\ n\geq 1\}$ weakly converges to a Gaussian random variable with zero mean and covariance (φ,φ). Therefore, the energy of the system without an interaction is obtained using the generalized Gaussian random element ξ in H.

Suppose that the particles can interact with one another. Let $f : \mathcal{X}^k \to R$ be a bounded, symmetric and measurable function. The random variable

$$\int_0^T f(x_1(t),\ldots,x_k(t))dt = L(f,x_1,\ldots,x_k)$$

can be treated as the energy of an interaction of the k particles during $[0;T]$. Denote, by $\mathcal{M}_k$, the family of such f for which

$$E\left\{\int_0^T f(x_1(t),\ldots,x_k(t))dt/x_1(s),\ldots,x_{k-1}(s), s\in[0;T]\right\} = 0.$$

The completion of $\mathcal{M}_k$ in the norm

$$\|f\|_k^2 := EL(f,x_1,\ldots,x_k)^2$$

coincides with the symmetric part of the k-th tensor of H and is denoted by H_k. The following statement can be proved similarly to theorem 4.2.1 (cf. [35]).

Theorem 4.3.1. For every square-integrable functional η from ξ,

$$\eta = \sum_{k=0}^{-} f_k(\xi,\cdots,\xi),$$

the correspondent energy of the interaction of the particles

$$W_N = \sum_{k=0}^{N} \frac{1}{N^{\frac{k}{2}}} \sum_{t_1,\dots,t_k} L(f_k, x_{t_1}, \dots, x_{t_k}),$$

weakly converges to η when $N \to \infty$.

The random elements in H or H_k can be treated as potentials of interaction which depend on the particles.

Example 4.4.1. Consider the random element in H

$$\eta = \sum_{k=0}^{-} f_k(\cdot, \xi, .\overset{k}{.}., \xi).$$

η is a random potential depending on the system of particles. The extended stochastic integral from η (if it exists) is the weak limit of the sequence

$$\sum_{k=0}^{N} \frac{1}{N^{\frac{k}{2}}} \sum_{-t_1,\dots,t_k-} L(f_{k-1}, x_{t_1}, \dots, x_{t_k}), N \geq 1.$$

The elements of this sequence correspond to the energy of the new system , which arises after addition of the observed particle.

Chapter 5

Equations with random operators

5.1 Estimation of the moments of multilinear forms from white noise

Let, as in the preceding chapters, ξ be a generalized Gaussian random element in the real separable Hilbert space H with zero mean and identity covariance operator. Use the notation from section 1.1.

Theorem 5.1.1. Let $A_n \in H_n, n \geq 1$. Then

$\forall\, p \in N$:

$$E[A_n(\xi, \dots, \xi)]^{2p} \leq (n!)^{2p-1} \cdot p^{2(n+1)p} \cdot 2^{(n+2)p} \cdot \|A_n\|_n^{2p}.$$

Proof. Let $\{\varphi_k;\ k \geq 1\}$ be an orthonormal basis in H. Then

$$A_n = \sum_{k_1 \dots k_n = 1}^{\infty} a_{k_1 \dots k_n} \cdot \varphi_{k_1} \otimes \dots \otimes \varphi_{k_n}.$$

The distribution of $A_n(\xi, \dots, \xi)$ and, consequently, its moments depend only on the values $\{a_{k_1 \dots k_n};\ k_1 \dots k_n \geq 1\}$. Consider, in $L_2([0;1])$, the Rademaher system $\{f_k; k \geq 1\}$ and white noise $\tilde{\xi}$, which is defined using the Wiener process w.

$$(h, \tilde{\xi}) := \int_0^1 h(t) dw(t), h \in L_2([0;1]).$$

The random variable $F_n(\tilde{\xi}, \dots, \tilde{\xi})$, where

$$F_n = \sum_{k_1 \dots k_n = 1}^{\infty} a_{k_1} \dots a_{k_n} f_{k_1} \otimes \dots \otimes f_{k_n},$$

has the same distribution as $A_n(\xi, \dots, \xi)$. As in section 1.1

$$F_n(\tilde{\xi}, \dots, \tilde{\xi}) = \int_{\Delta_n(1)} f(t_1, \dots, t_n) dw(t_1) \dots dw(t_n), (5.1.1)$$

where

$$f(t_1,\ldots,t_n) = n! \sum_{k_1\ldots k_n=1} a_{k_1\ldots k_n} f_{k_1}(t_1)\cdot\ldots\cdot f_{k_n}(t_n),$$

the series converges in $L_2([0;1]^n)$. Note that, for every $p \in N$, $f \in L_{2p}([0;1]^n)$. Indeed, since the functions $\{f_k; k \geq 1\}$ are independent random variables on $[0;1]$ with Lebesgue measure,

$$\frac{1}{n!^{2p}} \int_0^1 \overset{n}{\cdots} \int_0^1 f(t_1,\ldots,t_n)^{2p} dt_1 \ldots dt_n =$$

$$= \sum_{i_1+\ldots+i_p=pi_1,\ldots,i_p-0} C_{2p}^{2i_1,\ldots,2i_p} \prod_{s=1}^{p} \sum_{k=1\ldots k_n=1}^{-} a_{k_1\ldots k_n}^{2i_s} \operatorname{sign}(i_s) \leq \|A_n\|_n^{2p} \cdot 4^p \cdot p^{2p}.$$

Apply to the integral (5.1.1) the estimate for the moments o f the stochastic integrals from [20].

$$E\Big\{\int_0^1 \int_0^{t_n} \cdots \int_0^{t_2} f(t_1,\ldots,t_n) dw(t_1)\ldots dw(t_n)\Big\}^{2p} \leq$$

$$\leq [p(2p-1)]^p \cdot \int_0^1 E\Big\{\int_0^{t_n} \cdots \int_0^{t_2} f(t_1,\ldots,t_n) dw(t_1)\ldots dw(t_{n-1})\Big\}^{2p} dt_n \leq \ldots \leq$$

$$\leq [p(2p-1)]^{np} \cdot \int_{\Delta_n(1)} f(t_1,\ldots,t_n)^{2p} dt_1 \ldots dt_n \leq$$

$$\leq \|A_n\|_n^{2p} \cdot 4^p \cdot p^{2p} \cdot [p(2p-1)]^{pn} \cdot (n!)^{2p-1}.$$

Theorem is proved.

Consequence.

$$E[A_n(\xi,\ldots,\xi)]^{2p} \leq (n!)^{p-1} \cdot 2^{(n+2)p} \cdot p^{2(n+1)p} \cdot [EA_n^2(\xi,\ldots,\xi)]^p.$$

This inequality follows from the preceding theorem and the property

$$EA_n^2(\xi,\ldots,\xi) = n! \cdot \|A_n\|_n^2.$$

Consider the case $A_n \in (H_n, \mathcal{H})$, where $\mathcal{H}$ is the real separable Hilbert space. Let $\{e_k;\ k \geq 1\}$ be an orthonormal basis in $\mathcal{H}$. Then

$$E|A_n(\xi,\ldots,\xi)|^{2p} = E\left(\sum_{k=1}^{-} [A_n(\xi,\ldots,\xi), e_k]^2\right)^p =$$

$$= \sum_{k_1\ldots k_{2p}=1}^{-} E \prod_{i=1}^{p} [A_n(\xi,\ldots,\xi), e_{k_i}]^2 \leq$$

$$\leq \sum_{k_1\ldots k_{2p}=1}^{-} \prod_{i=1}^{p} \left(E[A_n(\xi,\ldots,\xi), e_{k_i}]^{2^{i+1}}\right)^{\frac{1}{2^i}} \leq$$

$$\leq \sum_{k_1\ldots k_p=1}^{-} \prod_{i=1}^{p} \left((n!)^{2^i-1} \cdot 2^{(n+2)2^i} \cdot 2^{i2(n+1)-2^i} \cdot (E[A_n(\xi,\ldots,\xi), e_{k_i}]^2)^{2^i}\right)^{\frac{1}{2^i}} \leq$$

$$\leq (n!)^p \cdot 2^{n+2} \cdot 2^{p^2(n+1)} \cdot \left(E|A_n(\xi,\ldots,\xi)|^2\right)^p.$$

Theorem 5.1.1 gives the following sufficient condition for the random process to be continuous with probability 1. Call the process $\{\eta(t);\ t \in [0;1]\}$ polynomial if there exists $N \geq 1$ such that, for every $t \in [0;1]$, $\eta(t)$ is a polynomial from ξ (see chapter 3) of degree not greater then N.

Theorem 5.1.2. A polynomial random process, which satisfies the Gelder condition in the square-mean, is continuous with probability 1.

Proof. Consider a polynomial process η, for which

$$\forall\, t,s \in [0;1]:$$

$$E|\eta(t) - \eta(s)|^2 \leq C \cdot |t-s|^\alpha, \tag{5.1.2}$$

where $\alpha \in (0;1)$. Denote by N the maximum degree of the Hilbert-Schmidt form from ξ, which occurs in the expansion of η.

Then

$$\forall\, t \in [0;1]:$$

$$\eta(t) = \sum_{k=0} A_k(t,\xi,\ldots,\xi).$$

Due to its orthogonality, the summands of the expansion of η satisfy condition (5.1.2). Take $p \in N$ such that $p\alpha > 1$. Then, from theorem 5.1.1,

$$\forall\, k = 1,\ldots,N:\ \forall\, t,s \in [0;1]:$$

$$E\ |A_k(t,\xi,\ldots,\xi) - A_k(s,\xi,\ldots,\xi)|^{2p} \leq$$

$$\leq C_1 \cdot \left(E|A_k(t,\xi,\ldots,\xi) - A_k(s,\xi,\ldots,\xi)|^2\right)^p \leq C_1 \cdot C^p \cdot |t-s|^{p\alpha}.$$

It follows from the Kolmogorov theorem [22] that, for every $k = 1,\ldots,N$, the random process $\{A_k(t,\xi,\ldots,\xi);\ t \in [0;1]\}$ satisfies, with probability 1, the Gelder condition, with index $(p\alpha - 1)\ (2p)$. Since p can be chosen at will, then the process $\{A_k(t,\xi,\ldots,\xi); t \in [0;1]\}$ and the process η satisfy, with probability 1, the Gelder condition, with an arbitrary index from $(0;\alpha/2)$. The theorem is proved.

5.2 Equations with random operators under moment conditions

This section is devoted to the following two types of equations:

$$x + A(x) = y, \tag{5.2.1}$$

$$\frac{dx(t)}{dt} = A(x(t)), x(0) = y, \tag{5.2.2}$$

where A is a random operator in H and y is the known random element. It will be supposed that all random elements have finite second moment of the norm, and definition 3.3.1 will be used without the localization for the action of the random operator on the random elements. Under such conditions, the main instrument for the solution of (5.2.1) and (5.2.2) is the application of the following statement.

Lemma 5.2.1. Let GSRO A be jointly Gaussian with ξ and $(E+MA)^{-1} \in L(H)$ exist, where E is the identity operator in H. Then (5.2.1) has no more than one solution.

Proof. It is enough to check that the equation

$$x + Ax = 0 \tag{5.2.3}$$

has exactly one solution $x = 0$. Let x be an arbitrary solution of (5.2.3). Consider the Ito-Wiener expansion of x

$$x = \sum_{k=0}^{\infty} S_k(\xi, \dots, \xi),$$

and A

$$\forall\, \varphi \in H : A(\varphi) = a_0(\varphi) + a_1(\varphi)(\xi),$$

where $a_0 = MA$. It follows from the definition of the action of A on x, that

$$S_0 + a_0(S_0) = 0,$$

$$S_1 + a_1(S_0) + \Lambda a_0(S_1) = 0 \in H_1,$$

$$S_2 + \Lambda a_1(S_1) + \Lambda a_0(S_2) = 0 \in H_2,$$

$$\vdots$$

$$S_n + \Lambda a_1(S_{n-1}) + \Lambda a_0(S_n) = 0 \in H_n,$$

$$\vdots$$

Consequently,

$$\forall\, n \geq 0 : S_n = 0,$$

i.e. $x = 0$. Lemma is proved.

Suppose that, for the GSRO A, $\mathbf{E}A = 0$. Under the condition $(E+\mathbf{E}A)^{-1} \in L(H)$ it is sufficient to consider only the such case for (5.2.1). Indeed, the equations

$$x + A(x) = y$$

and

$$x + (E + MA)^{-1}(A - MA)(x) = (E + MA)^{-1}(y)$$

are equivalent. In (5.2.2) only the random part of A plays a role.

Lemma 5.2.2. Let A be GSRO with zero-mean and jointly Gaussian with $\xi, y = y_0 \in H$. Then (5.2.1) has a solution if, and only if, the series

$$\sum_{n=0}^{\infty}(-1)^n A^n(y_0)$$

converges in $L_2(\Omega, P, H)$, and the solution is equal to the sum of this series.

This statement can be easily checked using the Ito–Wiener expansion.

To study the case of the random right part in (5.2.1), consider the functions from the Ornstain-Ulenbek operator $N = I(D)$, which was defined in the preceding chapter. Let $\{r_n; n \geq 0\}$ be the sequence of the positive numbers. Consider the following self-adjoint operator in $L_2(\Omega, F, P)$. For the random variable α with the Ito–Wiener expansion

$$\alpha = \sum_{n=0}^{\infty} S_n(\xi, \ldots, \xi), \quad N_r\alpha = \sum_{n=0}^{\infty} r_n S_n(\xi, \ldots, \xi),$$

if this series converges in $L_2(\Omega, F, P)$. In the particular case, $r_n = n$, the operator N_r coincides with the Ornstain-Ulenbek operator N. For further consideration some properties of N_r are needed. It has been proved, in chapter 1, that the operators D, I and their combinations are local. For an arbitrary sequence $\{r_n; n \geq 0\}$, N_r is not local (in general). For example, the following statement is true.

Lemma 5.2.3. If the operator N_r is local and bounded then $r_n = r_m$ for $n, m \geq 0$.

Proof. It is enough to check that $N_r - r_0 E = 0$. Consider the contraction of N_r on the subspace R of $L_2(\Omega, F, P)$, which consists of functions of the kind $f((\varphi, \xi))$, where $\varphi \in H$ – is a fixed element and $\|\varphi\| = 1$. N_r maps R onto R. The subspace R can be treated as $L_2(R, \mu)$, where μ is a Gaussian measure on R with zero-mean and variance 1. N_r is a bounded and local operator in $L_2(R, \mu)$. Hence N_r is a multiplication operator in $L_2(R, \mu)$. Since $N_r(1) = r_0$ then

$$\forall h \in L_2(R, \mu) : N_r(h) = r_0 \cdot h.$$

But, from consideration of the lemma,

$$\forall h = \sum_{n=0}^{\infty} \alpha_n \overline{H}((\varphi, \xi));$$

$$N_r(h) = \sum_{n=0}^{\infty} r_n \alpha_n \overline{H}((\varphi, \xi)),$$

and $S_0, r_n = r_0, n \geq 0$. Lemma is proved.

Note that the property of locality determines the structure of N_r not only in the 'bounded' case. It can be proved that, if the sequence $\{r_n; n \geq 0\}$ has polynomial growth, the local operator N_r is a polynomial from the Ornstein–Ulenbek operator $I(D)$. An analogous statement for an operator in $L_2([0; 2\pi])$ can be found in [36]. The operator N_r also can be defined not only for random variables but for random elements in H, similar to (5.2.4).

To determine the operator N_r which can be used for solving equation (5.2.1), consider the following forms on H:

$$\forall m \geq 2 \quad \forall u, \varphi_1, \ldots, \varphi_m \in H :$$
$$a_m(u; \varphi_1, \ldots, \varphi_m) = a_1(\ldots a_1(a_1(u; \varphi_1); \varphi_2) \ldots ; \varphi_m).$$

For each $m \geq 1$, a_m can be treated as a bounded linear operator from H onto the space of m-linear Hilbert–Schmidt forms on H. Denote by $\|a_m\|$ the operator norm of this operator.

Theorem 5.2.1. Let $\forall k \geq 0, m \geq 0$:

$$\|a_m\| \cdot \frac{\sqrt{(k+m)!}}{\sqrt{k!}} \leq c_m r_k,$$

where

$$\sum_{m=0}^{\infty} c_m < +\infty, \quad \forall l \geq 1: \lim_{k\to\infty} \frac{k^l}{r_k} = 0.$$

Then equation (5.2.1) has a solution for every $y \in \mathcal{D}(N_r)$.

Proof. Consider $y \in \mathcal{D}(N_r)$

$$y = \sum_{k=0}^{\infty} S_k(\xi, \dots, \xi).$$

It follows from the definition that, for $m \geq 1$

$$A^m(y) = \sum_{k=0}^{\infty} \Lambda a_m(S_k)(\xi, \overset{k+m}{\dots}, \xi).$$

This series converges in $L_2(\Omega, P, H)$ because

$$\sum_{k=0}^{\infty} (k+m)!\|\Lambda a_m(S_k)\|^2_{k+m} \leq$$

$$\sum_{k=0}^{\infty} \frac{\|a_m\|^2(k+m)!}{k!} \cdot k!\|S_k\|^2_k \leq$$

$$\leq \sum_{k=0}^{\infty} c_m^2 r_k^2 k!\|S_k\|^2_k = c_m^2 M\|N_r(y)\|^2.$$

Hence

$$\sum_{m=0}^{\infty} \sqrt{M\|A^m(y)\|^2} \leq \sum_{m=0}^{\infty} c_m \sqrt{M\|N_r(y)\|^2} \leq +\infty.$$

So the Neiman series

$$x = \sum_{m=0}^{\infty} (-1)^m A^m(y)$$

converges in $L_2(\Omega, P, H)$. Since the operator A is closed (cf. section 3.1), x is the solution of (5.2.1).

In the case $H = L_2([0,1])$ the question of the smoothness of the trajectories of the solution can be discussed.

Theorem 5.2.2. Let, for the certain orthonormal basis $\varphi_k; k \geq 1$ in $L_2([0,1])$,

1) $\forall k \geq 1 \ \forall h \in C([0;1]):$
   $a_1(h)(\varphi_k) \in C([0;1]);$

2) $\exists\, K > 0: \ \forall s,t \in [0,1] \ \forall h \in C([0;1]):$
   $\sum_{k=0}^{\infty}(a_1(h)(\varphi_k)(t) - a_1(h)(\varphi_k)(s))^2 \leq K \cdot \|h\|^2 \cdot |t-s|;$

3) $\exists\, C > 0 \ \forall n \geq 1: \ \|\Lambda a_n(y_0)\|^2_n \leq \frac{C^n}{(n!)^2}.$

Then the equation

$$x + A(x) = y_0$$

has the unique solution x and, if $y_0 \in Lip^{\frac{1}{2}}([0;1])$, for every $\beta \in (0;\frac{1}{2}) x \in Lip^{\beta}([0;1])$ with probability 1.

Proof. The existence of the unique solution x follows from lemmas 5.2.1 and 5.2.2. Consider $y_0 \in Lip^{\frac{1}{2}}([0;1])$. Verify firstly that, for every $n \geq 1$, $A^n(y_0) \in Lip^{\beta}([0;1])$ with probability 1 for $\beta \in (0;1/2)$. Denote

$$\xi_k := (\varphi_k;\xi), k \geq 1.$$

$\{\xi_k; k \geq 1\}$ is the sequence of independent identically distributed Gaussian variables with zero-mean and variance 1. The value $A^n(y_0)$ can be written as follows

$$\Lambda^n(y_0) = \sum_{k_1,\ldots,k_n=1}^{\infty} \Lambda a_n(y_0)(\varphi_{k_1},\ldots,\varphi_{k_n})\xi_{k_1} * \ldots * \xi_{k_n},$$

where $\xi_{k_1} * \ldots * \xi_{k_n}$ as in the section 1.1 is a multiple Hermit polynomial. Denote, for $N \geq 1$,

$$S_N = \sum_{k_1,\ldots,k_n=1}^{N} \Lambda a_n(y_0)(\varphi_{k_1},\ldots,\varphi_{k_n})\xi_{k_1} * \ldots * \xi_{k_n} =$$

$$= \sum_{k_1,\ldots,k_n=1}^{N} a(\Lambda a_{n-1}(y_0)(\varphi_{k_1},\ldots,\varphi_{k_{n-1}}))(\varphi_{k_n}) \cdot \xi_{k_1} * \ldots * \xi_{k_n}.$$

It follows, from 1), that S_N is a random element in $C([0;1])$ for every $N \geq 1$. To prove that $\{S_N; N \geq 1\}$ converges in $C([0;1])$ in probability, it is sufficient [37] to check the following conditions.

I. For a certain set X which is dense in $[0;1]$, for every $t \in X$, the sequence $\{S_N(t); N \geq 1\}$ converges in probability.

II. $\exists\, C_1, \sigma, \delta > 0 : \ \forall t, s \in [0,1]:$

$$\sup_{N \geq 1} M|S_N(t) - S_N(s)|^{\sigma} \leq C_1|t-s|^{1+\delta}.$$

Condition I can be checked as follows. For every $t \in [0;1]$, the sequence $\{S_N(t); N \geq 1\}$ is a martingale with respect to the flow $\{\sigma(\xi_1,\ldots,\xi_N); N \geq 1\}$ and

$$M\ S_N(t)^2 \leq n! \sum_{k_1,\ldots,k_n=1}^{N} a_n(y_0)(\varphi_{k_1},\ldots,\varphi_{k_n})^2(t).$$

Since

$$\int_0^1 \sum_{k_1,\ldots,k_n=1}^{\infty} a_n(y_0)(\varphi_{k_1},\ldots,\varphi_{k_n})^2(t)dt =$$

$$= \|a_n(y_0)\|_n^2 < +\infty,$$

then, for almost all $t \in [0;1]$ with respect to the Lebesgue measure,

$$\sup_{N \geq 1} M S_N(t)^2 < +\infty.$$

So the sequence $\{S_N(t); N \geq 1\}$ converges in probability for almost all $t \in [0;1]$. To verify condition II consider, for every $p \in \mathcal{N}$,

$$M(S_N(t) - S_N(s))^{2p} =$$

$$= M(\sum_{k_1,\dots,k_n=1}^{N} a(\Lambda a_{n-1}(y_0)(\varphi_{k_1},\dots,\varphi_{k_{n-1}}))(\varphi_{k_n})(t) -$$

$$-a(\Lambda a_{n-1}(y_0)(\varphi_{k_1},\dots,\varphi_{k_{n-1}}))(\varphi_{k_n})(s)]\xi_{k_1} * \cdot * \xi_{k_n})^{2p} =$$

$$\sum_{l=1}^{2p} \sum_{\substack{r_1+\dots+r_l=2p \\ r_1,\dots,r_l\geq 1 \\ \vec{k_j}\neq\vec{k_i}, i\neq j}} E\frac{(2p)!}{r_1!\cdot\ldots\cdot r_l!}[\Delta a_{n,\vec{k}_1}]^{r_1}\times$$

$$\times\xi_{\vec{k}_1}^{r_1}\cdot\ldots\cdot[\Delta a_{n,\vec{k}_l}]^{r_l}\cdot\xi_{\vec{k}_l}^{r_l},$$

where $\vec{k} = (k_1,\dots,k_n) \in \{1,\dots,N\}^n$,

$$\Delta a_{n,\vec{k}} := a(\Lambda a_{n-1}(y_0)(\varphi_{k_1},\dots,\varphi_{k_{n-1}}))(\varphi_{k_n})(t) -$$

$$-a(\Lambda a_{n-1}(y_0)(\varphi_{k_1},\dots,\varphi_{k_{n-1}}))(\varphi_{k_n})(s),$$

$$\xi_{\vec{k}} := \xi_{k_1} * \cdot * \xi_{k_n}.$$

Firstly, consider

$$E\sum_{k_1\dots k_n=1}^{N}(\Delta a_{n,\vec{k}})^r\cdot|\xi_{\vec{k}}|^r.$$

It follows from theorem 5.1.1 and the equality

$$\overline{H}_n(w(1)) = n!\int_{\Delta_n(1)} 1dw(t_1)\dots dw(t_n),$$

that, for the Gaussian random variable η with zero-mean and variance 1,

$$E\overline{H}_n(\eta)^{2m} \leq (n!)^{2m-1}(m[2m-1])^{m-n}.$$

So for $r = 2m, m \geq 1$,

$$E\sum_{k_1\dots k_n=1}^{N}(\Delta a_{n,\vec{k}})^{2m}\cdot|\xi_{\vec{k}}|^{2m} \leq (n!)^{2m-1}(Q_m)^n\times$$

$$\times\left[\sum_{k_1\dots k_n=1}^{N}(a(\Lambda a_{n-1}(y_0)(\varphi_{k_1},\dots,\varphi_{k_{n-1}}))(\varphi_{k_n})(s))^2\right]^m \leq$$

$$\leq (n!)^{2m-1}\cdot(Q_m)^n\cdot\left[\sum_{k_1\dots k_{n-1}=1}^{N}\|\Delta\Lambda a_{n-1}(y_0)(\varphi_{k_1},\dots,\varphi_{k_{n-1}})\|^2\cdot K\cdot|t-s|\right]^m \leq$$

$$\leq (Q_m)^n\cdot K^m\cdot\frac{C^{m(n-1)}}{(n-1)!^{2m}}\cdot(n!)^{2m-1}\cdot|t-s|^m \leq$$

$$\le K^m (Q_m)^n \cdot C^{m(n-1)} \cdot n^{2m-1} \cdot \frac{|t-s|^m}{(n-1)!}.$$

Since from Holder's inequality for certain $\lambda_p > 0$, $R_p > 0$ which depend only from p,

$$E|S_N(t) - S_N(s)|^{2p} \le |t-s|^p \cdot R_p^n \cdot n^{2p} \frac{1}{(n!)^{\lambda_p}} \tag{5.2.4}$$

Consequently the sequence $\{S_N;\ N \ge 1\}$ converges in probability in $C([0;1])$. So

$$A^n(y_0) \in C([0;1]) \ (\text{mod P}), n \ge 1.$$

It follows from (5.2.4) that, for every $p \ge 1$, the sequence $\{|S_N(t) - S_N(s)|^2 p;\ N \ge 1\}$ is uniformly integrable under t, s being fixed. Hence

$$E|A_n(y_0)(t) - A_n(y_0)(s)|^{2p} \le$$

$$\le |t-s|^p \cdot R_p^n \cdot n^{2p} \cdot \frac{1}{(n!)^{\lambda_p}}.$$

Prove that the series $\sum_{n=1}^{\infty} A_n(y_0)$ converges in $C([0;1])$ in probability. From the above

$$E \sum_{n=1}^{\infty} \|A^n(y_0)\|^2 < +\infty,$$

i.e.

$$\int_0^1 E \sum_{n=1}^{\infty} [A^n(y_0)(t)]^2 dt < +\infty.$$

Consequently, for almost all $t \in [0;1]$

$$E \sum_{n=1}^{\infty} [A^n(y_0)(t)]^2 < +\infty.$$

Since for $m \ne n$

$$EA^m(y_0)(t) \cdot A^n(y_0)(t) = 0,$$

then the series

$$\sum_{n=1}^{\infty} (-1)^n A^n(y_0)(t)$$

converges in $L_2(\Omega, \mathcal{F}, P)$ (hence in probability) for almost all $t \in [0;1]$. It follows from (5.2.4) and Holders inequality that condition **II** holds. It can be proved similarly to (5.4.3) that

$$\forall\, p \in N : \exists C_p > 0 :\ \forall\, t, s \in [0;1] :$$

$$E\Big(\sum_{n=1}^{\infty} (-1)^n A^n(y_0)(t) - \sum_{n=1}^{\infty} (-1)^n A^n(y_0)(s)\Big)^{2p} \le C_p \cdot |t-s|.$$

Hence, due to the Kolmogorov theorem, for the every $\beta \in (0;1/2)$

$$\sum_{n=1}^{\infty} (-1)^n A^n(y_0) \in \text{Lip}^{\beta}([0;1]) (\text{mod P}).$$

Theorem is proved.

Note that the statement of the theorem can be easily formulated for the case $EA \neq 0$.

Consider an example.

Example 5.2.1. Let $\{Q_k;\ k \geq 1\}$ be the sequence of bounded linear operators in $L_2([0;1])$ such that

1) $\forall\, k_1, \dots, k_n \geq 1\ \forall\, x \in L_2([0;1])$:

$$\|Q_{k_1} \dots Q_{k_n}(x)\| \leq \frac{1}{n!}\|x\|;$$

2) $\forall\, k \geq 1\ \forall\, x \in L_2([0;1])$:

$$Q_k(x) \in C([0;1]);$$

3) $\exists\, C > 0\ \forall\, x \in L_2([0;1])$:

$$\sup_{k\geq 1} \sup_{t\neq s} \frac{1}{t-s}(Q_k(x)(t) - Q_k(x)(s))^2 \leq C \cdot \|x\|^2.$$

To build GSRO A, which satisfies the condition of theorem 5.2.1, use the sequence $\{Q_k;\ k \geq 1\}$. Sequence $\{Q_k;\ k \geq 1\}$ will be considered independently in future. Let $\{\varphi_k;\ k \geq 1\}$ be an orthonormal basis in $L_2([0;1])$ and $\{\alpha_k;\ k \geq 1\}$ be the real-valued sequence from l_2 . Define the operator A in such a way

$$\forall\, x \in L_2([0;1]) :$$

$$A(x) = \sum_{k=1}^{\infty} \alpha_k(\varphi_k; \xi) \cdot Q_k(x),$$

where the series co converges in the square-mean. The bilinear form, which corresponds to A has a following representation

$$\forall\, x, \varphi \in L_2([0;1]) :$$

$$a(x)(\varphi) = \sum_{k=1}^{\infty} \alpha_k \cdot (\varphi_k; \varphi) \cdot Q_k(x).$$

It can be easily checked that the form a and the basis

$$\{\varphi_k; k \geq 1\}$$

satisfy the conditions of theorem 5.2.1. Build the sequence $\{Q_k; k \geq 1\}$ now. Consider in the space $B := \mathrm{Lip}^{\frac{1}{2}}([0;1])$ the following norm:

$$\forall\, f \in B: \ \|f\|_L := |f(0)| + \sup_{t\neq s} \frac{|f(t) - f(s)|}{|t-s|^{\frac{1}{2}}}.$$

With norm $\|\cdot\|_L$ the space B is Banach. B is dense in $L_2([0;1])$. Hence $L_2([0;1])$ can be developed in the infinite orthogonal sum of subspaces $\{X_k;\ k \geq 1\}$ in such a manner that

$$\forall\, k \geq 1: \ \dim X_k \cap B = \infty.$$

Denote, for every $k \geq 1$, $Y_k := X_k \cap B$ be the closed subspace of B. Let $U_k : X_k \to Y_k$ be a bounded linear operator with norm 1, $k \geq 1$.

Let $\{b_{ij}; i, j \geq 1\}$ be such matrix, that

$$\forall\, i, j \geq 1 : \ |b_{ij} < \frac{1}{j}.$$

Define, for every $n \geq 1$,

$$\forall\, x \in L_2([0; 1]) : \ Q_n(x) := \sum_{k=1}^{\infty} b_{kn} U_k P_k(x),$$

where P is a projector on $X_k, k \geq 1$. It can be easily checked that $\{Q_n; n \geq 1\}$ satisfies conditions 1) – 3).

5.3 Boundary value problems

Consider the boundary value problem

$$L_t x(t) + \partial_t^* K_t x(t) = y(t), t \in (0; 1),$$

$$u_i(x) = 0, i = 1, \ldots, l, \tag{5.3.1}$$

where $L_t x(t) = x^{(n)}(t) + a_1(t) x^{(n-1)}(t) + \cdots + a_n(t) x(t)$ and $K_t x(t) = x^{(m)}(t) + b_1(t) x^{(m-1)}(t) + \cdots + b_m(t) x(t)$ are linear differential expressions in an unknown random function $\{x(t); t \in [0; 1]\}$ with non-random coefficients $a_i \in C^n([0; 1]), i = 1, \ldots n$, $b_j \in C^m([0; 1]), j = 1, \ldots m$, $l = \max(m, n)$ and $U_i, i = 1, \ldots, l$ are boundary conditions linear in x [38]. The unknown random process $\{x(t); t \in [0; 1]\}$ is given on the probability space $(\Omega, \mathcal{F}, P)$ and is assumed to be measurable with respect to the σ-algebra generated by the Wiener process $\{w(t); t \in [0; 1]\}$. The derivatives of x with respect to t are defined by the mean-square convergence and ∂_t^* is the operator adjoint to the operator of stochastic differentiation with respect to $w(t)$ (see chapter 3). The random process $\{y(t); t \in [0; 1]\}$ is also assumed to be measurable with respect to the σ-algebra generated by the process $\{w(t); t \in [0; 1]\}$ which satisfies the condition $Ey(t)^2 < +\infty$, for all $t \in [0; 1]$, and is mean-square continuous.

Definition 5.3.1. A random process $\{x(t); t \in [0; 1]\}$ is called a solution of problem (5.3.1) if $Ex(t)^2 < +\infty$ for each $t \in [0; 1]$, the value $\partial_t^* x(t)$ is defined and equalities (5.3.1) hold with probability 1.

Note that, for $n = 1$ and $m = 0$, problem (5.3.1) is a linear stochastic differential equation with a Cauchy condition at 0 or 1. Because of the boundary conditions, the solution of (5.3.1), if it exists, is not, in general, a random process that is non-anticipating with respect to the flow of σ-algebras generated by the Wiener process. In order to consider this problem, the exponential transformation of the Wiener functionals can be used. To solve the problem it is sufficient to find the exponential transformation of the process $\{x(t); t \in [0; 1]\}$. For $\varphi \in C^\infty([0; 1])$ denote

$$X(\varphi, t) = \tau(x(t))(\varphi) =$$

$$= \exp \frac{1}{2} \int_0^1 \varphi^2(\tau) d\tau \cdot E \exp \left\{ i \int_0^1 \varphi(\tau) dw(\tau) \right\} \cdot x(t)$$

$$Y(\varphi,t)=\tau(y(t))(\varphi)=$$
$$=\exp\frac{1}{2}\int_0^1\varphi^2(\tau)d\tau\cdot E\exp\left\{i\int_0^1\varphi(\tau)dw(\tau)\right\}\cdot y(t).$$

From (3.5.1) the following family of boundary value problems dependent on $\varphi\in C^\infty([0;1])$ as a parameter is obtained:

$$\begin{cases} L_t\ \ X(\varphi,t)+i\varphi(t)\quad K_tX(\varphi,t)=Y(\varphi,t), \\ \quad U_i(X(\varphi,\cdot))=0,\quad i=1,\dots,l. \end{cases}\tag{5.3.2}$$

Here the fact that the exponential transformation carries the operator ∂_t^* into the operator of multiplication by $i\varphi(t)$ has been used. Denote by $V_1(\varphi,\cdot),\dots,V_l(\varphi,\cdot)$ a fundamental system of solutions of the linear differential equation

$$L_tV(\varphi,t)+i\varphi(t)K_tV(\varphi,t)=0,$$

that satisfies the initial conditions

$$V_i^{(j-1)}(\varphi,0)=\delta_{ij},i,j=1,\dots,l.$$

Let $W(\varphi,t)$ be the Wronskian determinant of the system $V_1(\varphi,\cdot),\dots,\ V_l(\varphi,\cdot)$ for $t>0$. Define

$$g(\varphi,t,s)=\pm\frac{1}{2W(\varphi,s)}\begin{vmatrix} V_1(\varphi,t) & \dots & V_l(\varphi,t) \\ V_1^{(l-2)}(\varphi,s) & \dots & V_l^{(l-2)}(\varphi,s) \\ & \vdots & \\ V_1(\varphi,s) & \dots & V_l(\varphi,s) \end{vmatrix},$$

where the sign " + " is used for $t\geq s$ and " − " is used for $t<s$.

Let

$$\Delta(\varphi)=\begin{vmatrix} U_1(V_1) & \dots & U_1(V_l) \\ & \vdots & \\ U_l(V_1) & \dots & U_l(V_l) \end{vmatrix},$$

$$H(\varphi,t,s)=\begin{vmatrix} V_1(\varphi,t) & \dots & V_l(\varphi,t) & g(\varphi,t,s) \\ U_1(V_1) & \dots & U_1(V_l) & U_1(g) \\ & \vdots & & \\ U_l(V_1) & \dots & U_l(V_l) & U_l(g) \end{vmatrix}.$$

Now the following statements about the problem (5.3.1) can be formulated.

Theorem 5.3.1. Suppose that $n>m+1$ and the usual boundary value problem

$$L_tu(t)=0,\ U_i(u)=0,\ i=1,\dots,n,$$

has only the one solution $u\equiv 0$. Then problem (5.3.1) has at most one solution.

Theorem 5.3.2. If m=0 then problem (5.3.1) has, at most, one solution.

Theorem 5.3.3. Suppose that the condition of theorem 5.3.1 holds. If the exponential transform $\{Y(\varphi,t);t\in[0;1]\}$ of the right part $\{y(\varphi,t);t\in[0;1]\}$ has the form $Y(\varphi,t)=\Delta(\varphi)\cdot\Gamma(\varphi,t);t\in[0;1]$, where

$$\Gamma(\varphi,t)=g_0(t)+\sum_{n=1}^{\infty}i^n\int_{\Delta_n(1)}g_n(t,\tau_1,\dots,\tau_n)\prod_{i=1}^n\varphi(\tau_i)d\vec{\tau},$$

and the kernels $\{g_n; n \geq 0\}$ are such that there is a $C > 0$ for which

$$\forall n \geq 0 : \text{esssup}\{|g_n(t, \tau_1, \ldots, \tau_n)| : t \in [0; 1], 0 \leq \tau_1 \leq \ldots \leq \tau_n \leq 1\} \leq C^n,$$

then problem (5.3.1) has a unique solution $\{x(t); t \in [0; 1]\}$ satisfying with probability 1 the Lipschitz condition with an arbitrary exponent $\alpha < \frac{1}{2}$.

Consider the proofs.

Proof of theorem 5.3.1. It suffices to show that under the conditions of the theorem the problem

$$\begin{cases} L_t x(t) + \partial_t^* K_t x(t) = 0, \\ U_i(x) = 0, i = 1, \ldots, n \end{cases}$$

has only the identically zero solution. Let $\{V(\varphi, t); t \in [0; 1]\}$ be the exponential transform of the solution of problem (5.3.2). Fix $\varphi \in D$ and $t \in [0; 1]$ and consider the function $R \ni z \to V(z\varphi, t) = \alpha(z)$ of a real variable z. According to the definition of the exponential transformation, α is an analytic function on R. Consider another function obtained similarly: $R \ni z \to \Delta(z\varphi) = \beta(z)$. It follows, from the construction of α and β, that $\{z : \alpha(z) \neq 0\} \subset \{z : \beta(z) = 0\}$ [38]. Since $\beta(0) \neq 0$ is a consideration and β is analytic on R, the set $\{z : \beta(z) = 0\}$ is at most countable. Consequently, since α is analytic, $\alpha(z) = 0$ for all $z \in R$. In particular, $0 = \alpha(1) = V(\varphi, t)$. Since φ and t were chosen arbitrarily, $V(\varphi, t) = 0$ for all $\varphi \in D$ and $t \in [0; 1]$. Consequently, $x(t) = 0 \ (\text{mod} P), t \in [0; 1]$. Theorem is proved.

Proof of theorem 5.3.2. Suppose, as in the proof of theorem 5.3.1, that $\{V(\varphi, t); t \in [0; 1]\}$ is the exponential transformation of the solution of the homogeneous problem. Consider the function $R \ni z \to V(\varphi + (-i)zI, t) = \alpha(z)$ of a real variable z, where I is the function identically equal to 1 on [0;1]; α is an analytic function on R that can be non-zero only at points of the spectrum of the differential operator corresponding to the expression $L_t \cdot +i\varphi(t)$ and to the boundary conditions $U_i, i = 1, \ldots, n$. According to [38] such a spectrum consists of, at most, denumerable set of isolated points. Therefore, $\alpha(z) = 0$ for all $z \in R$. In particular, $0 = \alpha(0) = V(\varphi, t)$. Since φ and t are arbitrary, the solution of the homogeneous problem is identically zero. Theorem is proved.

Proof of theorem 5.3.3. Under the conditions of the theorem, problem (5.3.2) has a solution and it is given by

$$X(\varphi, t) = (-1)^n \int_0^1 H(\varphi, t, s)\Gamma(\varphi, s)ds$$

for each φ. To prove that $X(\varphi, t)$ is the result of the exponential transformation of some random variable and to verify the sample continuity of the random process $\{x(t); t \in [0; 1]\}$ $X(\varphi, t)$ can be represented as the sum of the series of multiple integrals of φ. Use the explicit expression of H. First consider the function g in this expression. Since $n > m + 1$ by the condition of the theorem, the Wronskian determinant of the fundamental system $V_1(\varphi, \cdot), \ldots, V_l(\varphi, \cdot)$ does not depend on φ,

$$W(s) = \exp\left(-\int_0^s a_1(\tau)d\tau\right), \quad s \in [0; 1].$$

It is easy to show that the solutions $\{V_i(\varphi), i = 1, \ldots, n\}$ themselves can be represented in the form

$$V_i(\varphi, t) = e_{i0}(t) + \sum_{m=1}^{\infty} \int_{\Delta_m(t)} e_{i_m}(t, \tau_1, \ldots, \tau_m) \prod_{j=1}^{m} \varphi(\tau_j) d\vec{\tau}, \ i = 1, \ldots, n. \quad (5.3.4)$$

Further, the kernels e_{im}, $i = 1, \ldots, n$ and $m \geq 0$, are differentiable with respect to t, and there exists a $C_1 > 0$ such that

$$|e_{im}(t, \tau_1, \ldots, \tau_m)| + |\frac{\partial}{\partial t} e_{im}(t, \tau_1, \ldots, \tau_m)| \leq C_1^m$$

$$\forall\, i = 1, \ldots, n, \ m \geq 0, \ 0 \leq \tau_1 \leq \ldots \leq \tau_m \leq t \leq 1.$$

The product $V_{i_1}(\varphi, t), \cdot \ldots \cdot V_{i_n}(\varphi, t)(\varphi, t)$ can also be represented like (4) and the corresponding kernels also satisfy the growth restrictions, but with another constant C_2. Consequently, the solution $X(\varphi, t)$ can be written as follows:

$$X(\varphi, t) = \sum_{i=1}^{n} V_i(\varphi, t) \left\{ f_{i0}(t) + \sum_{k=1}^{\infty} \int_{\Delta_k(1)} f_{ik}(t, \tau_1, \ldots, \tau_k) \cdot \prod_{[i=1)}^{k} \varphi(\tau_i) d\vec{\tau} \right\} \quad (5.3.5)$$

where the kernels, $f_{ik}, i = 1, \ldots, n$, for $k \geq 0$, satisfy a growth restriction analogous to the condition for kernels e_{ik} with a constant C_3 and a Lipschitz condition with respect to t with constant C_4^k for the f_{ik}, $i = 1, \ldots, n$.

It follows from (5.3.5) that $X(\varphi, t)$ is the result of the exponential transformation of some random variable $x(t) \in L_2(\Omega, \mathcal{F}, P)$ that is measurable with respect to the process $\{w(t); t \in [0; 1]\}$. To prove the sample continuity of the process $\{x(t); t \in [0; 1]\}$ note that

$$x(t) = \sum_{i=1}^{n} \eta_i(t),$$

where the random variables $\eta_i(t), i = 1, \ldots, n$ are carried into the individual terms of (5.3.5) under the exponential transformation. It can be proved similarly to section 5.2 that for each $i = 1, \ldots, n$ the process $\{\eta_i(t); t \in [0; 1]\}$ satisfies with probability 1 a Lipschitz condition with arbitrary exponent $\alpha < \frac{1}{2}$. Theorem is proved.

5.4 Stochastic calculus and partial differential equations of the first order

Consider the following example.

Example 5.4.1. Let $H = R$ and the white noise ξ be a Gaussian random variable with zero mean and variance 1. Consider, for the certain random variable α, the following equation

$$x - I(x) = \alpha. \quad (5.4.1)$$

According to chapter 1, (5.4.1) can be written

$$f(t) - tf(t) - f'(t) = g(t), t \in R, \quad (5.4.2)$$

where $x = f(\xi), \alpha = g(\xi)$. For the continuous g, equations (5.4.2) and (5.4.1) have infinitely many solutions from $\tilde{D}$. In certain cases this set of solutions contains square-integrable solutions. In other cases (5.4.1) may have the generalized solution, as in chapter 3.

The aim of this section is to determine a solution of the stochastic integral equation which is a usual stochastic process and a generalized functional of the first kind simultaneously.

Now consider the equation

$$x(t) + \int_0^1 a(t,s)x(s)ds + \int_0^1 b(t,s)x(s)dw(s) = y(t),\ t \in [0;1], \tag{5.4.3}$$

where $\{x(t); t \in [0;1]\}$, $\{y(t); t \in [0;1]\}$ are unknown and known random processes respectively. The kernels a and b are the deterministic step functions on $[0;1]^2$ and the localized extended stochastic integral is used in the left part. Since kernels a and b are the step functions, the certain solutions of (5.4.3) can be obtained by consideration of the following system

$$x_k + \sum_{j=1}^N a_{kj}x_j + \sum_{j=1}^N b_{kj}\left(x_j\xi_j - \frac{\partial x_j}{\partial \xi_j}\right) = y_k, k = 1,\dots,N, \tag{5.4.4}$$

where $\xi_1,\dots,\xi_N$ are independent Gaussian variables with zero mean and variance 1, which correspond to the increments of the process w; coefficients a_{kj}, b_{kj} correspond to the values of a and b; the random variables $\{x_k;\ k = 1,\dots,N\}$, $\{y_k;\ k = 1,\dots,N\}$ are measurable with respect to $\xi_1,\dots,\xi_N$ and, for the unknown x_j, the value $x_j\xi_j - \frac{\partial x_j}{\partial \xi_j}$ is the value of the extended stochastic integral in this situation (cf. chapter 1).

Theorem 5.4.1. Let the right parts of (5.4.4) satisfy the following conditions:

1) $\forall\ k = 1,\dots,N, \forall\ l = 0,1,\dots N, \forall\ t_1,\dots,t_l \in \{1,\dots,N\}$

$$\exists : y_k \cdot \xi_{t_1} * \dots * \xi_{t_l} :\in L_2(\Omega, \mathcal{F}, P),$$

where $\xi_{t_1} * \dots * \xi_{t_l}$ is defined as in chapter 1 and $: \cdot :$ is the Wick product (i.e. the extended stochastic integral in the finite-dimensional case);

2) The exponential transformation of the random variables from 1) have bounded sequences of derivatives at the zero point.

Then the system has the solution which is the regular functional of the first kind from $\xi_1,\dots,\xi_N$. If the matrix

$$(\delta_{kj} + a_{kj})_{kj=1}^N$$

is invertible, then the regular solution is unique.

Proof. Prove the existence of the solution. Denote

$$\forall\ k = 1,\dots,N :$$

$$\tilde{x}_k = \exp(-\frac{1}{2}\|\xi\|^2) \cdot x_k,$$

$$\tilde{y}_k = \exp(-\frac{1}{2}\|\xi\|^2) \cdot y_k,$$

where $\xi = (\xi_1, \ldots, \xi_N) and\ ||\xi||^2 = \sum_{k=1}^N \xi_k^2$. The system can be rewritten as follows:

$$\tilde{x}_k + \sum_{j=1}^N a_{kj}\tilde{x}_k - \sum_{j=1}^N b_{kj}\frac{\partial \tilde{x}_j}{\partial \xi_j} = \tilde{y}_k, \quad k = 1, \ldots, N.$$

This system is a system of linear equations in partial derivative form with constant coefficients, which can be written in the form

$$(\delta_{kj} + a_{kj} - b_{kj}\frac{\partial}{\partial \xi_j})_{kj=1}^N \ \tilde{x} = \tilde{y}, \tag{5.4.5}$$

where $\tilde{x} = (\tilde{x}_1, \ldots, \tilde{x}_N) and \tilde{y} = (\tilde{y}_1, \ldots, \tilde{y}_N)$. Denote by $P(\frac{\partial}{\partial \xi_1}, \ldots, \frac{\partial}{\partial \xi_N})$ the differential operator-determinant of the matrix from (5.4.5), and by $(Q_{kj})_{kj=1}^N$ the matrix from algebraic supplements of this matrix. Note, that the action of the differential operators Q_{kj} on the coordinates $\tilde{y}$ is defined using Sobolev derivatives in the space $L_2(R^N, \exp(-\frac{1}{2}||\xi||^2)d\xi)$. For example

$$\frac{\partial}{\partial \xi_1}\tilde{y}_j = -\exp(-\frac{1}{2}||\xi||^2)(y_j\xi_1 - \frac{\partial y_j}{\partial \xi_1}) = -\exp(-\frac{1}{2}||\xi||^2) \cdot : y_j \cdot \xi_1 :$$

If the functions $\tilde{x}_1, \ldots, \tilde{x}_N$ satisfy the equations

$$(P(\tilde{x}_1), \ \ldots \ , P(\tilde{x}_N)) = (Q_{kj})_{kj=1}^N \tilde{y}, \tag{5.4.6}$$

then $\tilde{x}_1, \ldots, \tilde{x}_N$ are the solutions of the system. Denote

$$\tilde{z} = (Q_{kj})_{kj=1}^N \tilde{y}.$$

Note that

$$\tilde{z} = \exp(-\frac{1}{2}||\xi||^2) \cdot (z_1, \ \ldots \ , z_N),$$

where, due to the conditions of the theorem, $z_1, \ \ldots \ , z_N \in L_2$ and the exponential transformations $\gamma_1, \ \ldots \ , \gamma_N$ of $z_1, \ \ldots \ , z_N$ satisfy condition 2). Let $\tilde{P}(\lambda) = P(i\lambda_1, \ \ldots \ , i\lambda_N)$. Consider the regular generalised functionals of the first kind $x_1, \ \ldots \ , x_N$, such that

$$T(x_j)(\lambda) = \frac{\gamma_j(\lambda)}{\tilde{P}(\lambda)},$$

where T is the symbol of the exponential transformation. The functions $\tilde{x}_1, \ \ldots \ , \tilde{x}_N$ obtained from $x_1, \ \ldots \ , x_N$ satisfy (5.4.6). So the random variables $x_1, \ \ldots \ , x_N$ are the solution of (5.4.5). Note that $x_1, \ \ldots \ , x_N$ may have not finite second moment and the Wick products $: x_j \cdot \xi_j :$ are defined by the localisation.

Now prove the uniqueness of the regular solution under the invertible matrix $(\delta_{kj} + a_{kj})_{kj=1}^N$. It follows, from this condition, that the coefficients of the formal Ito-Wiener expansions of $x_1, \ \ldots \ , x_N$ are uniquely determined by the right part. Let $j \in \{1, \ldots, N\}$ and values $\lambda_1^\circ, \ \ldots \ , \lambda_N^\circ \in C$ be fixed. For the function of the complex variable ζ

$$T(x_j)(\lambda_1^\circ \zeta, \ \ldots \ , \lambda_N^\circ \zeta)$$

the Taylor expansion, in the neighbourhood of zero, is known. So if $x'_1, \ \ldots \ , x'_N$ are the other regular solutions of (5.4.4) then, in the certain neighbourhood of zero,

$$T(x_j)(\lambda_1^\circ \zeta, \ \ldots \ , \lambda_N^\circ \zeta) = T(x'_j)(\lambda_1^\circ \zeta, \ \ldots \ , \lambda_N^\circ \zeta).$$

Let

$$T(x_j)(\lambda) = \frac{\gamma_1(\lambda)}{R_1(\lambda)}, \quad T(x'_j)(\lambda) = \frac{\gamma_2(\lambda)}{R_2(\lambda)}.$$

Then, in the neighbourhood of zero, and consequently on all C,

$$\gamma_1(\lambda_1^\circ\zeta, \; \dots \; , \lambda_N^\circ\zeta) R_2(\lambda_1^\circ\zeta, \; \dots \; , \lambda_N^\circ\zeta) = \gamma_2(\lambda_1^\circ\zeta, \; \dots \; , \lambda_N^\circ\zeta) R_1(\lambda_1^\circ\zeta, \; \dots \; , \lambda_N^\circ\zeta)$$

So there exist polynomials Q_1, Q_2, Q_3 analytical on all C functions $\tilde{\gamma}$ such that

$$R_1(\lambda_1^0\zeta, \dots, \lambda_N^0\zeta) = Q_1(\zeta) \cdot Q_2(\zeta),$$

$$R_2(\lambda_1^0\zeta, \dots, \lambda_N^0\zeta) = Q_2(\zeta) \cdot Q_3(\zeta),$$

$$\gamma_1(\lambda_1^0\zeta, \dots, \lambda_N^0\zeta) = \tilde{\gamma}(\zeta) \cdot Q_1(\zeta),$$

$$\gamma_2(\lambda_1^0\zeta, \dots, \lambda_N^0\zeta) = \tilde{\gamma}(\zeta) \cdot Q_3(\zeta).$$

Consequently, for all ζ such that $Q_2(\zeta) \neq 0$ the following equality holds

$$\frac{\gamma_1(\lambda_1^0\zeta, \dots, \lambda_N^0\zeta)}{R_1(\lambda_1^0\zeta, \dots, \lambda_N^0\zeta)} = \frac{\tilde{\gamma}(\zeta)}{Q_2(\zeta)} = \frac{\gamma_2(\lambda_1^0\zeta, \dots, \lambda_N^0\zeta)}{R_2(\lambda_1^0\zeta, \dots, \lambda_N^0\zeta)}$$

Since the values $\lambda_1^0, \dots, \lambda_N^0$ are arbitrary, then the functions γ_1/R_1 and γ_2/R_2 coincide throughout the domain of definition. So, in (3.5.8) the Hermander steps can be chosen in the same way for γ_1/R_1 and for γ_2/R_2. Consequently $x_j = x'_j$. The theorem is proved.

Continue example 5.4.1.

Example 5.4.1′. The solution of (5.4.1) for $\alpha = 1$ can be obtained as the generalized Wiener functional

$$x = \sum_{k=0}^{\infty} H_k(\xi_1).$$

Its T-transformation is

$$T(\lambda) = \sum_{k=0}^{\infty} (i\lambda)^k = \frac{1}{1 - i\lambda}, \quad |\lambda| < 1.$$

Consequently x is a regular functional. To obtain x take the real line of Hermander steps for the polynomial $P(\lambda) = 1 - i\lambda$. Then

$$x(t) = \frac{1}{\sqrt{2\pi}} \int_{-\infty}^{+\infty} \frac{e^{-s^2/2}}{1 - is} e^{-its} ds e^{t^2/2} =$$

$$= \frac{1}{\sqrt{2\pi}} \int_{-\infty}^{+\infty} \frac{e^{-s^2/2}}{1 + s^2} (\cos(ts) + s \sin(ts)) ds e^{t^2/2}.$$

So the solution of (5.4.1) is a regular generalised functional of the form

$$x = \frac{1}{\sqrt{2\pi}} \int_{-\infty}^{+\infty} \frac{e^{-s^2/2}}{1 + s^2} (\cos(s \cdot \xi_1) + s \sin(s \cdot \xi_1)) ds e^{\xi_1{}^2/2}.$$

To study equations with general kernels, the following result can be useful.

Theorem 5.4.2. Let $\{L_n; n \geq 1\}$ be the sequence of measurable bounded kernels on $[0;1]^2 and \{y_n; n \geq 1\}$ be the sequence of the right parts for which there exist the solutions $\{x_n; n \geq 1\}$ of the equations

$$x_n(t) + \int_0^1 L_n(t,s)x_n(s)dw(s) = y_n(t), \quad t \in [0;1], \ n \geq 1.$$

Suppose that there exists a sequence of smooth open sets $\{G_m; m \geq 1\}$,increasing to Ω, which are generated by double-differentiable random variables such that

$$1) \ \forall m \geq 1: \int_{G_m}\int_0^1 (x_n(t) - x_0(t))^2 dtdP \to 0, \quad n \to \infty,$$

$$\int_{G_m}\int_0^1 (y_n(t) - y_0(t))^2 dtdP \to 0, \quad n \to \infty;$$

$$2) \ \int_0^1\int_0^1 (L_n(t,s) - L_0(t,s))^2 dtds \to 0, \quad n \to \infty.$$

Than

$$x_0(t) + \int_0^1 L_0(t,s)x_0(s)dw(s) = y_0(t), \quad t \in [0;1].$$

Proof. First consider the case when $G_1 = G_2 = \ldots = \Omega$,

$$\forall n \geq 0: \quad M\int_0^1 x_n^2(t)dt < +\infty,$$

and the extended stochastic integral is defined using the Ito–Wiener expansion. Let

$$x_n(t) = \sum_{k=0}^{\infty} \int_{\Delta_k} p_{kn}(t,\tau_1,\ldots,\tau_k)dw(\tau_1)\ldots dw(\tau_k),$$

$$y_n(t) = \sum_{k=0}^{\infty} \int_{\Delta_k} q_{kn}(t,\tau_1,\ldots,\tau_k)dw(\tau_1)\ldots dw(\tau_k),$$

where $n \geq 0$ for the corresponding Ito -Wiener expansions. Then $\forall k \geq 0:$

$$\lim_{n\to\infty}\int_0^1\int_{\Delta_k} (p_{kn}(t,\tau_1,\ldots,\tau_k) - p_{k0}(t,\tau_1,\ldots,\tau_k))^2 d\tau_1\ldots d\tau_k dt = 0,$$

$$\lim_{n\to\infty}\int_0^1\int_{\Delta_k} (q_{kn}(t,\tau_1,\ldots,\tau_k) - q_{k0}(t,\tau_1,\ldots,\tau_k))^2 d\tau_1\ldots d\tau_k dt = 0.$$

It follows, from the uniqueness of the Ito -Wiener expansion, that $\forall k \geq 1: \ \forall n \geq 1:$

$$\int_{\Delta_k} p_{kn}(t,\tau_1,\ldots,\tau_k)dw(\tau_1)\ldots dw(\tau_k) +$$

$$+\int_0^1 L_n(t,s)\int_{\Delta_{k-1}} p_{k-1n}(s,\tau_1,\ldots,\tau_{k-1})dw(\tau_1)\ldots dw(\tau_{k-1})dw(s) =$$

$$= \int_{\Delta_k} q_{kn}(t, \tau_1, \dots, \tau_k) dw(\tau_1) \dots dw(\tau_k).$$

Consequently the same equalities hold for $n = 0$. So the extended stochastic integral

$$\int_0^1 L_0(t, s) x_0(s) dw(s)$$

is well-defined for almost all $t \in [0; 1]$ and

$$x_0(t) + \int_0^1 L_0(t, s) x_0(s) dw(s) = y_0(t), t \in [0; 1].$$

Consider the general case. For every $m \geq 1$ build the random variable $\alpha_m \in W^1$ such that

1) $\{|\alpha_m| > 0\} \subset\subset G_m$;
2) $|\alpha_m| \leq 1, \|D\alpha\| \leq 2 \pmod P$;
3) $\exists G'_m \in \mathcal{E}, G'_m \subset \{\alpha_m = 1\}$,

$$\cup_{m=1}^{\infty} G'_m = \Omega \pmod P.$$

The sequences $\{\alpha_m x_n;\ n \geq 1\}$ and $\{\alpha_m y_n;\ n \geq 1\}$ converge in L_2-sense on all Ω. For the random process $\alpha_m x_n$ the following equality holds

$$\alpha_m x_n(t) + \int_0^1 L_n(t, s) \alpha_m x_n(s) dw(s) =$$

$$= \alpha_m x_n(t) + \alpha_m \cdot \int_0^1 L_n(t, s) x_n(s) dw(s) -$$

$$- \int_0^1 L_n(t, s) x_n(s) D\alpha_m(s) dw(s), t \in [0; 1].$$

Consequently

$$\alpha_m x_n(t) + \int_0^1 L_n(t, s) \alpha_m x_n(s) dw(s) =$$

$$= \alpha_m y_n(t) - \int_0^1 L_n(t, s) x_n(s) \cdot D\alpha(s) ds.$$

As has already been proved for convergence in L_2-sense on all Ω, the random process $\alpha_m x_0$ satisfies the equality

$$\alpha_m x_0(t) + \int_0^1 L_0(t, s) \alpha_m x_0(s) dw(s) =$$

$$= \alpha_m y_0(t) - \int_0^1 L_0(t, s) x_0(s) \cdot D\alpha(s) ds, t \in [0; 1].$$

Since $\alpha_m \equiv 1$ and $D\alpha_m \equiv 0$ on G'_m then x_0 locally belongs to the domain of the definition of the stochastic integral operator

$$\int_0^1 L_0(t, s) dw(s), t \in [0; 1],$$

and the following equality holds

$$x_0(t) + \int_0^1 L_0(t,s)x_0(s)dw(s) = y_0(t), t \in [0;1].$$

The theorem is proved.

Example 5.4.2. Consider the equation

$$x(t) + \int_0^1 L_0(t,s)x(s)dw(s) = 1, t \in [0;1], \tag{5.4.7}$$

where

$$L_0(t,s) = \mathcal{X}_{\{s\le t\}}(t,s) + \sum_{r=1}^{k} \mathcal{X}_{\Delta_r}(t)\cdot \mathcal{X}_{\Delta_r}(s)\cdot \mathcal{X}_{\{s>t\}}(t,s)$$

and $\Delta_r, r = 1,\dots,k$ are disjoint intervals from $[0;1]$. Choose the sequence $\{L_n; n \ge 1\}$ in the following way:

$$L_n(t,s) = \sum_{1\le j_2<j_1\le n} \mathcal{X}_{[\frac{j_1-1}{n};\frac{j_1}{n}]}(t)\cdot \mathcal{X}_{[\frac{j_2-1}{n};\frac{j_2}{n}]}(s) +$$

$$+\sum_{r=1}^{k}\sum_{j_1 j_2=a_{rn}}^{b_{rn}} \mathcal{X}_{[\frac{j_1-1}{n};\frac{j_1}{n}]}(t)\cdot \mathcal{X}_{[\frac{j_2-1}{n};\frac{j_2}{n}]}(s)\cdot \mathcal{X}_{\{j_2\ge j_1\}}(j_1,j_2),$$

where the values a_{rn}, b_{rn}, $r = 1,\dots,k$ are such that

$$m\left(\left[\frac{a_{rn}}{n},\frac{b_{rn}}{n}\right] \triangle \Delta_r\right) \to 0, n\to\infty$$

where m is the Lebesgue measure. Take n sufficiently large for the intervals $\left[\frac{a_{rn}}{n},\frac{b_{rn}}{n}\right], r = 1,\dots,k$ to be disjoint. The equation with L_n can be considered finite-dimensional if

$$x_n(t) = \sum_{j=1}^{n} \mathcal{X}_{[\frac{j-1}{n},\frac{j}{n}]}(t)\cdot x_{jn},$$

where the random variables $\{x_{jn}; j = 1,\dots,n\}$ are the functions from $\{\xi_{jn} = \sqrt{n}\left(w(\frac{j}{n}) - w(\frac{j-1}{n})\right); j = 1,\dots,n\}$. An equation of type (5.4.4) is obtained:

$$x_{j_1 n} + \sum_{j_2=1} j_1 - 1\frac{1}{\sqrt{n}} : x_{j_2 n}\xi_{j_2 n} : + \tag{5.4.8}$$

$$+\sum_{r=1}^{k}\sum_{j_2=a_{rn}}^{b_{rn}} \mathcal{X}_{[\frac{a_{rn}}{n},\frac{b_{rn}}{n}]}\left(\frac{j_1-\frac{1}{2}}{n}\right)\cdot\frac{1}{\sqrt{n}}\cdot : x_{j_2 n}\cdot\xi_{j_2 n} : = 1, j_1 = 1,\dots,n.$$

It follows from theorem (5.4.1) that this equation has a regular solution. Consider

$$P_n(\lambda_1,\dots,\lambda_n) =$$

$$\det\left(\frac{1}{\sqrt{n}}\cdot\left(\mathcal{X}_{\{j_2<j_1\}}(j_1,j_2) + \sum_{r=1}^{k}\mathcal{X}_{[\frac{a_{rn}}{n},\frac{b_{rn}}{n}]}\left(\frac{j_1-\frac{1}{2}}{n}\right).\right.\right.$$

$$\cdot \mathcal{X}_{\left[\frac{a_{rn}}{n}, \frac{b_{rn}}{n}\right]}\left(\frac{j_1 - \frac{1}{2}}{n}\right) \mathcal{X}_{\{j_2<j_1\}}(j_1, j_2)\Big) \cdot i\lambda_{j_2} +$$

$$+\delta_{j_1 j_2})_{j_1 j_2=1}^{n} = \prod_{r=1}^{k}\left(\sum_{j=a_{rn}+1}^{b_{rn}} i\lambda_j + 1\right).$$

For every $\lambda_1, \ldots, \lambda_n \in R$:

$$|P_n(\lambda_1, \ldots, \lambda_n)| \geq 1.$$

So the Hermander steps can be chosen as a real subspace of C^n. Consequently the regular solution

$$x_{jn} = \frac{1}{\sqrt{2\pi}^n} \cdot \exp\left(\frac{1}{2}\sum_{j_1=1}^{n} \xi_{j_1 n}^2\right) \cdot \int_{R^n} \exp\left(-\frac{1}{2} \cdot \sum_{j_1=1}^{n} \lambda_{j_1}^2\right) \cdot$$

$$\cdot \bar{x}_{jn} \exp\left(-i\sum_{j_1=1}^{n} \lambda_{j_1} \cdot \xi_{j_1 n}\right) d\lambda_1 \ldots d\lambda_n, j = 1, \ldots, n,$$

where $\bar{x}_{jn}, j = 1, \ldots, n$ are found to be the solutions of the equation

$$A_n(\bar{x}_{1n}, \ldots, \bar{x}_{nn})^T = (1, \ldots, 1)^T.$$

For sufficiently large n

$$1 < a_{1n} < b_{1n} < a_{2n} < b_{2n} < \ldots < a_{kn} < b_{kn} < n.$$

Then

$$\bar{x}_{1n} = 1, \bar{x}_{2n} = 1 - i\frac{\lambda_1}{\sqrt{n}},$$

$$\bar{x}_{3n} = 1 - i\frac{\lambda_1}{\sqrt{n}} + i^2\frac{\lambda_1}{\sqrt{n}} \cdot \frac{\lambda_2}{\sqrt{n}},$$

$$\vdots$$

$$\bar{x}_{a_{1n}n} = 1 - \sum_{j_1=1}^{a_{1n}-1} i\frac{\lambda_{j_1}}{\sqrt{n}} + \sum_{1\leq j_1<j_2<a_{1n}} i\frac{\lambda_{j_1}}{\sqrt{n}} \cdot i\frac{\lambda_{j_2}}{\sqrt{n}} - \ldots +$$

$$+(-1)^{a_{1n}-1} \cdot i\frac{\lambda_1}{\sqrt{n}} \cdot i\frac{\lambda_2}{\sqrt{n}} \cdot \ldots \cdot i\frac{\lambda_{a_{1n}-1}}{\sqrt{n}},$$

$$\bar{x}_{a_{1n}+1n} = \bar{x}_{b_{1n}n} = \frac{1}{1+\sum_{j=a_{1n}+1}^{b_{1n}} i\frac{\lambda_j}{\sqrt{n}}} \cdot$$

$$\cdot\left(1 - \sum_{j_1=1}^{a_{1n}} i\frac{\lambda_{j_1}}{\sqrt{n}} + \sum_{1\leq j_1<j_2\leq a_{1n}} i\frac{\lambda_{j_1}}{\sqrt{n}} \cdot i\frac{\lambda_{j_2}}{\sqrt{n}} - \ldots\right.$$

$$\left.+(-1)^{a_{1n}} \cdot i\frac{\lambda_1}{\sqrt{n}} \cdot \ldots \cdot i\frac{\lambda_{a_{1n}}}{\sqrt{n}}\right),$$

$$\bar{x}_{b_{1n}+1n} = 1-(1-\bar{x}_{b_{1n}n}) = \bar{x}_{b_{1n}n},$$

$$\bar{x}_{b_{1n}+2n} = \bar{x}_{b_{1n}n} - \bar{x}_{b_{1n}n}\cdot i\frac{\lambda_{b_{1n}+1}}{\sqrt{n}},\ldots$$

Consequently

$$x_{1n}=1,\ x_{2n}=1-w(\frac{1}{n}),$$

$$x_{3n}=1-w(\frac{2}{n})+w(\frac{1}{n})\cdot(w(\frac{2}{n})-w(\frac{1}{n})),$$

$$\vdots$$

$$x_{a_{1n}n}=1-w\left(\frac{a_{1n}-1}{n}\right)+\sum_{1\le j_1<j_2<a_{1n}}\left(w(\frac{j_1}{n})-w(\frac{j_1-1}{n})\right)\times$$

$$\times\left(w(\frac{j_2}{n})-w(\frac{j_2-1}{n})\right)-\ldots+(-1)^{a_{1n}-1}w(\frac{1}{n})\times$$

$$\times\left(w(\frac{2}{n})-w(\frac{1}{n})\right)\cdot\left(w(\frac{a_{1n}-1}{n})-w(\frac{a_{1n}-2}{n})\right),$$

$$x_{a_{1n}+1n}=\ldots=x_{b_{1n}n}=\Big(1-w(\frac{a_{1n}}{n})+$$

$$+\sum_{1\le j_1<j_2\le a_{1n}}\left(w(\frac{j_1}{n})-w(\frac{j_1-1}{n})\right)\cdot\left(w(\frac{j_2}{n})-w(\frac{j_2-1}{n})\right)-$$

$$\ldots+(-1)^{a_{1n}}w(\frac{1}{n}\cdot\ldots\cdot\left(w(\frac{a_{1n}}{n})-w(\frac{a_{1n}-1}{n})\right)\Big)\cdot$$

$$\cdot f_{1n}\left(\frac{w(\frac{b_{1n}}{n})-w(\frac{a_{1n}}{n})}{\sqrt{\frac{b_{1n}-a_{1n}}{n}}}\right),$$

$$x_{b_{1n}+1n}=x_{b_{1n}n},$$

$$x_{b_{1n}+2n}=x_{b_{1n}n}\cdot\left(1-\left(w(\frac{b_{1n}+1}{n})-w(\frac{b_{1n}}{n})\right)\right),\ldots.$$

where the functions $f_{1n},\ldots,f_{kn}$, as in example 5.4.1, are as follows:

$$f_{rn}(t)=e^{\frac{t^2}{2}}\cdot\frac{1}{\sqrt{2\pi}}\cdot\int_{-\infty}^{+\infty}\frac{e^{-\frac{u^2}{2}}}{1+i\sqrt{\frac{b_{rn}-a_{rn}}{n}}\cdot u}e^{-itu}du, r=1,\ldots,k.$$

The sequence of solutions satisfies the conditions of theorem 5.4.2 with the sequence of smooth open sets

$$G_m=\{\omega:\ \max_{[0;1]}|w|<m\}.$$

The limit process has the form

$$x(t) = \prod_{a_r \le t} f_r\left(\frac{w(b_r) - w(a_r)}{\sqrt{b_r - a_r}}\right) \cdot \exp\left(-w(a_1) - \frac{a_1}{2}\right) \cdot$$

$$\exp\left(-w(a_2) - w(b_1)\frac{a_2 - b_1}{2}\right) \cdot \ldots \cdot \exp\left(-w(t) + w(b_{r_0}) - \frac{t - b_{r_0}}{2}\right),$$

where $[a_r; b_r] = \Delta_r, r = 1, \ldots, kandb_{r_0} \le t \le a_{r_0+1}$. This is the solution of the equation.

Note that the statement of theorem 5.4.2 remains true when the kernel of the deterministic integral is non-zero.

Bibliography

[1] H. -H. Kuo. *Gaussian measures in Banach spaces.* Lect. Notes in Math., **463** Berlin (1975)

[2] R. H. Cameron, W. T. Martin. *The orthogonal development of non-linear functionals in series of Fourier-Hermite functions.* Ann. of Math. Vol. 48, 385–392 (1947)

[3] V. G. Mazia. *Sobolev spaces.* Ed. in the Leningrad Univ., Leningrad (1985)

[4] H. Sugita. *Positive Wiener functionals and potential theory over abstract Wiener spaces.* Osaca Journ. Math. **25**, 3, 665–698

[5] E. Pardoux, D. Nualart. *Stochastic calculus with anticipating integrands.* Probability theory and related fields. **78,** 535–581 (1988)

[6] N. Danford, D. T. Shvarc. *Linear operators.* Vol. 1. *General theory.* Moscow (1962)

[7] Yu. L. Daletsky, S. N. Paramonova. *One formula of theory of Gaussian measures and estimate of stochastic integrals.* Probability theory and appl. **4,** 845–849 (1974)

[8] O. Enchev. *Stochastic integration with respect to Gaussian random measures.* Ph. D. Thesis, Sofia Univ., Sofia (1983)

[9] A. Yu. Shevliakov. *Ito formula for the extended stochastic integral.* Theory probability and math. statist.. 22, p. 163–174 (1981)

[10] A. A. Dorogovtsev. *Stochastic integrals with respect to Gaussian random measures.* Theory probability and math. statist.. **44,** p.53–59 (1992)

[11] S. Ogawa, T. Sekiguchi. *On the Ito formular of non-causal type.* Proc. Japan Acad. Ser. A. Math. Sci.. **60,** p.249–251 (1984)

[12] S. Ogawa. *Quelques proprietes de l'integrale stochastique du type non-causal.* Jap. J. Appl. Math.. **1,** p.405–416 (1984)

[13] V. Matskyavichus. *Symmetric stochastic integral equations of the Fredholm type.* Lithuanian Math. J.. **24,** p.121–130 (1984)

[14] E. Pardoux, P. Protter. *A two-sided stochastic integral and its calculus.* Probab. Theory and Related Fields. **78,** p.15–19 (1987)

[15] P. R. Halmos. Measure theory. Van Nostrand, Princeton, N.Y. (1950)

[16] S. Watanabe. *Stochastic differential equations and Malliavin calculus.* Tata institute of fundamental research. Bombay. (1984)

[17] A. A. Dorogovtsev. *The singular transformations of the Wiener functionals and two-sided stochastic integral of Pardoux-Protter.* Stochastic analysis and its Appl.. Kiev: Institute of Math.. p.37–44 (1989)

[18] A. V. Skorohod. Integration in Hilbert space. Nauka, Moscow (1975)

[19] R. Curant, D. Hilbert. Equations with partial derivatives. Mir, Moscow (1964)

[20] R. Sh. Liptser, A. N. Shiryaev. Statistics of random processes. Nauka, Moscow (1974)

[21] I. I. Gihman, A. V. Skorohod. The theory of random processes. Vol.1. Nauka, Moscow (1971)

[22] A. V. Skorokhod. Random linear operators. Naukova dumka, Kiev (1979)

[23] Yu. M. Berezansky. Self-adjoint operators in spaces of the functions of an infinite number of arguments. Naukova dumka, Kiev (1978)

[24] M. Sh. Birman, M. Z. Solomiak. The spectral theory of self-adjoint operators in Hilbert space. Leningrad university, Leningrad (1980)

[25] T. Sekiguchi, Y. Shiota. *L_2-theory of non-causal stochastic integrals.* Math. Rep. Toyama Univ.. **8**, p.119–195 (1985)

[26] Bharycha-Reid. Random integral equations. Acad. Press, New York (1972)

[27] T. Sekiguchi, Y. Shiota. *On a class of universally integrable random functions.* Tohoku math. Journ.. **38**, p.357–364 (1986)

[28] Yu. L. Daletsky, S. V. Fomin. Measures and differential equations in infinitely dimensional spaces. Nauka, Moscow (1983)

[29] H.-H. Kuo. *On Fourier transform of generalized Brownian functionals.* J. Multivariate Anal. **12**, p.415–431 (1982)

[30] L. Hormander. Linear differential operators with partial derivatives. Mir, Moscow (1965)

[31] N. N. Vahania, V. I. Tarieladze, S. A. Chobanian. *Probability distributions in Banach spaces.* Nauka, Moscow (1985)

[32] T. Hida. Brownian motion. Mir, Moscow (1989)

[33] F. A. Berezin. The method of second quantization. Nauka, Moscow (1986)

[34] Yu. V. Borovskih, V. S. Korolyuk. *The convergence of U-statistics to multiple Wiener integrals.* Reports of the Ac Sci of Ukraine. **4**, p.37–40 (1989)

[35] R. J. Adler, R. Epstein. *Some central limit theorems for Markov paths and some properties of Gaussian random fields.* Stochastic processes and their appl. Vol. 24, **2**, p.157–202 (1987)

[36] A. A. Dorogovtsev. *The local diagonal operators.* The Fourier series: theory and applications. Kiev: Inst. of Math. Ukr Ac Sci. p.27–30 (1992)

[37] V. V. Buldygin. Convergence of random elements in topological spaces. Naukova dumka, Kiev (1980)

[38] M. A. Naimark. Linear differential operators, 2nd ed.. Nauka, Moscow (1969)

Bei Fragen zur Produktsicherheit wenden Sie sich bitte an:
If you have any questions regarding product safety,
please contact:

Walter de Gruyter GmbH
Genthiner Straße 13
10785 Berlin
productsafety@degruyterbrill.com